Bibliografische Information der Deutschen Nationalbibliothek

Die Deutsche Nationalbibliothek verzeichnet diese Publikation in der
Deutschen Nationalbibliografie; detaillierte bibliografische Daten sind
im Internet über http://dnb.d-nb.de abrufbar.

ISBN 978-3-8325-2688-7

Logos Verlag Berlin GmbH
Comeniushof, Gubener Str. 47,
10243 Berlin
Tel.: +49 (0)30 42 85 10 90
Fax: +49 (0)30 42 85 10 92
INTERNET: http://www.logos-verlag.de

Rosa Maria Lo Presti

Geological vs. Climatological Diversification in the Mediterranean Area: Micro- and Macroevolutionary Approaches in *Anthemis* L. (Compositae, Anthemideae)

λογος

Diese Veröffentlichung wurde als Dissertation
bei der Universität Regensburg,
Institut für Botanik,
D-93053 Regensburg angefertigt
und am 03.11.2009 eingereicht.

E-mail: rmlopresti@gmail.com

Contents

Chapter 1

General Introduction

Biodiversity and evolution

The five mediterranean-climate regions of the world occupy less than 5% of the Earth´s surface but harbour ca. 20% of the world´s total plant species (Cowling & al., 1996a). These regions are characterized by exceptionally high levels of endemism (Myers & al., 2000) that are even higher than those of some tropical areas such as Peru or Panama (Goldblatt & Manning, 2002). Their prominent role as reservoirs for plant diversity has lead to their inclusion into the 34 global biodiversity hotspots (Myers, 1988; Myers & al., 2000; Mittermeier & al., 2005), i.e. those regions of the world that are featuring extraordinary species richness. During the past centuries, scientists have discovered, collected and described new species around the world, steadily improved the state of knowledge about global biodiversity and warned against its loss in a world threatened with anthropogenic habitat destruction and environmental pollution. But it is only since Darwin's *The Origin of Species* that a new, fundamental question has challenged the biologists: *why* are there so many species? Regions characterized by high species richness appear to be particularly suitable to investigate the evolutionary origins of diversity, or, in Darwin´s terms, the rise and multiplication of new species.

A review on patterns and determinants of plant species richness in the five mediterranean-climate regions of the world (Cowling & al., 1996a) has shown that on a global scale, climate change together with the incidence of fire could be considered the driving forces for explaining the evolution of diversity in these regions. In the Cape Floristic Region (CFR) and in south western Australia, the two regions characterized by the highest diversity at a regional scale (Cowling &

al., 1996a; Myers & al., 2000), high fire frequencies and relatively mild Quaternary glaciations have resulted in rapid speciation coupled with low extinction rates, which in turn have produced extremely species-rich areas. Harder glacial conditions in California and in the Mediterranean Basin have resulted in higher extinction rates, while the absence of fire in Chile has been related to the persistence of slow-evolving lineages (Cowling & al., 1996a). Moreover, while great diversity of nutritionally-impoverished soils has been invoked as a principle determinant of high diversity in Australia and the CFR (Cowling & al., 1996b; Hopper & Gioia, 2004), the role of tectonics in shaping the actual topographical heterogeneity of the Mediterranean Basin should not be forgotten (Thompson, 2005). As the vast amount of literature suggests, many species-level molecular phylogenies have been produced in the last decades to test the causes of radiations, such as *Nemesia* in the CFR (Scrophulariaceae; Datson & al., 2008), *Ranunculus* in the Mediterranean Basin (Ranunculaceae; Paun & al., 2005) and *Drosera* in Australia (Droseraceae; Yesson & Culham, 2006). While each study stresses different aspects, recognition of recent radiations (Late Miocene-Early Pliocene) and shift to the annual life form are common to all of them. These factors have mostly been related to another aspect common to all Mediterranean-climate regions, namely the adaptation to increasingly dry conditions culminating in the typical Mediterranean climate with its characteristic summer drought.

Speciation is primary the result of micro-evolutionary processes that prevent gene flow and promote population differentiation. Therefore, the study of the spatial structure of populations provides an important tool with which to investigate the evolutionary processes involved in the origin of new species. Phylogeographic studies conducted in the different mediterranean-climate regions of the world have shown that the flora of these regions is characterized by highly fragmented population systems (Coates, 2000). Endemic species, which are 'key ingredients' (Thompson, 2005) of the mediterranean-type flora, are also often featured with narrow geographical distributions (Linder, 2003; Hopper & Gioia, 2004; Thompson, 2005), and provide evidence for strong patterns of localized differentiation. Such patterns could be related to geographical isolation, which is considered an important factor limiting gene flow among populations (Grant, 1971; Linder 2003). Geographical isolation results from multiple causes ranging from sea-level fluctuations, which would disrupt continuous distribution ranges, to occasional dispersal across islands or intermontane valleys. The heterogeneity of the mediterranean-type regions provides ideal settings in which isolation could occur, and has been invoked, for example, to explain patterns of genetic diversity in populations of *Cytisus villosus* (Fabaceae) in Sicily and in the Aeolian islands (Troia & al., 1997). However, several features of plant population ecology and evolution suggest that this mode of speciation may be of secondary importance (Thompson, 2005), and that speciation in the mediterranean-type regions is more locally driven. The local speciation model is built upon two microevolutionary processes which promote population differentiation (Thompson, 2005): first, genetic drift and random differentiation provide novel variants; second, novel variants may be favoured by natural selection if they have adaptive advantages. Random loss and fixation of alleles in disjunct parts of a species range provide the template for the accumulation of genetic variation, which in turn can be subjected to disruptive selection in the presence of environmental gradients (e.g. climatic, topographic or edaphic). Such processes are very likely to occur in the heterogeneous landscapes of the Mediterranean regions as suggested by numerous examples of ecologically differentiated sister taxa: Edaphic specialization in the CFR was first demonstrated by Rourke (1972) for *Leucospermum* (Proteaceae) and later confirmed for many other taxa (see Linder, 2003 for a summary); in the Mediterranean Basin, the disjunct distribution of closely related taxa in *Cyclamen* (Primulaceae; Thompson & al., 2005), *Saxifraga* (Saxifragaceae; Conti & al., 1999) or *Senecio* (Asteraceae; Comes & Abbott, 2001), has been related to ecological gradients.

Among the five mediterranean-type regions of the world, the Mediterranean Basin is the biggest in terms of total area covered (60% of the world´s total mediterranean-type regions; Dallman, 1998) and it houses 10% of the world's total floristic richness (Médail & Quézel, 1997). Trapped in a collision zone between the African and Eurasian plates, the Mediterranean Sea represents the largest inland sea of the world and is decorated with islands which vary from tiny fragments of previous land-bridge connections to the big islands with their massive mountains and volcanoes.

The limits of the Mediterranean region were in the past defined on the basis of floristic or vegetation criteria (by using the distribution of particular species, such as *Olea europaea* L. or *Quercus ilex* L., or that of the sclerophyllous forest association), but it became soon clear that neither of them corresponded to the actual extension of the biogeographical or ecological Mediterranean region (Quézel, 1985). The iso-climatic area proposed by Daget (1977a, b) extends away from the Mediterranean Basin reaching into sub-Saharan Africa, Arabia and western Asia. More recent delimitations (Quézel & Médail, 2003; Thompson, 2005) cover the region where an effective drought occurs in the warmest part of the year, which is the critical defining characteristic of the Mediterranean region (see Fig. 1.1).

Biogeographically, the Mediterranean Basin includes 11 hotspots defined on the basis of plant endemism and richness (Fig. 1.1; Médail & Quézel, 1997; Véla & Benhouhou, 2007). It has been evidenced that both ecological specialization and geographical isolation have been primary determining factors to explain such a high biodiversity (Thompson, 2005). Tectonics provided a template for plant evolution, which has been further modulated by climate; spatial heterogeneity and a variety of substrates provided the basis of habitat mosaic, which is associated with local variation in selection pressure (Thompson, 2005). This region seems therefore particularly suitable as a model system in which to integrate the study of species divergence (macroevolution) with that of population differentiation (microevolution).

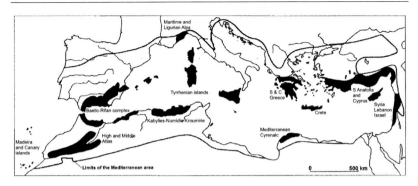

Fig. 1.1. The delimitation of the Mediterranean region, along with the 11 regional hotspots of plant biodiversity (from Médail & Quézel, 1997; Véla & Benhouhou, 2007).

The geological and climatological history of the Mediterranean

The formation and development of the Mediterranean region occurred mainly during the Cenozoic (Meulenkamp & Sissingh, 2003; Jolivet & al., 2006) as a result of the convergence of the Eurasian and African plates and their associated microplates. During the Palaeogene (until c. 20 Ma) the climatic conditions remained subtropical (Jolivet & al., 2006), allowing the continued presence of a tropical flora (Axelrod, 1975; Ivanov & al., 2002; Agusti & al., 2003; Thompson, 2005). From the Middle Miocene (c. 15 Ma) onwards, the transformation of the Mediterranean region into its land-locked position became more pronounced due to the closing of hitherto existing sea connections between the Paratethys and the Indian Ocean (c. 13 Ma), between the Paratethys and the Mediterranean Sea (c. 10 Ma), and finally between the Mediterranean Sea and the Indian Ocean (c. 8 Ma) (Jolivet & al., 2006). In parallel, the climatic conditions underwent considerable changes and developed, after repeated cooling events between 14 and 10 Ma (Van Dam, 2006), into a more temperate climate with seasonal contrasts in the temperature regime (Thompson, 2005).

A trend towards increasing aridification started in the whole Mediterranean region around 9-8 Ma (Ivanov & al., 2002; Fortelius & al., 2006; Van Dam,

2006) and was probably correlated with the northward shift of the Subtropical High Pressure Zone (Van Dam, 2006) and the closing of the Mediterranean Sea in the east (see Jolivet & al., 2006). Within this general trend, a west–east gradient of decreasing humidity was established around 11-9 Ma, followed by the formation of a north–south gradient (c. 7-5 Ma) with dry conditions in the south and relatively humid conditions in the north (Fortelius & al., 2006). This aridification process led to the progressive extinction of tropical elements, for example mangroves and Taxodiaceae forests, and to their substitution by sclerophyllous plant communities (Thompson, 2005), persisting into the Early Pliocene (c. 4 Ma) and being further accompanied by a progressive cooling (Fauquette & al., 1999; Van Dam, 2006). While the Messinian Salinity Crisis at the end of the Miocene (5.96-5.33 Ma; Hsü, 1972; Krijgsman, 2002) considerably changed the land–water distribution, there is little evidence that this event was accompanied by dramatic climatic changes (Suc, 1984; Fauquette & al., 2006). Therefore, the general trend towards aridification and cooling continued undiminished and led, with the onset of a marked seasonality and stable summer droughts, to the establishment of a Mediterranean type of climate around 3 Ma (Suc, 1984; Thompson, 2005). From the Late Pliocene (c. 2.5-2.1 Ma) onwards, no modification in the geodynamic situation was observed, while the gradual cooling and drying of the climate culminated in the onset of the glacial–interglacial alternation of the Quaternary (around 2.4 Ma; Bertoldi & al., 1989; Combourieu-Nebout, 1993; Thompson, 2005).

The genus *Anthemis* L.

In the Mediterranean area, the genus *Anthemis* L. (Asteraceae, Anthemideae) provides a suitable plant group with which to link both the macro- and the microevolutionary approaches. As it was reconstructed to have diverged from its close relatives between 10 to 15 Ma (Oberprieler, 2005), it may act as a suitable proxy for the reconstruction of the biogeographical and climatological history of the Mediterranean area, spanning the transition from the subtropical climate of the Early Miocene to the typical Mediterranean environment of the present. On the other side, it includes many closely related groups of species, such as the *A. secundiramea* group widespread across the Sicilian Channel, which provide suitable models to study the role of geographical and/or ecological diversification on a more local scale.

a)   b)

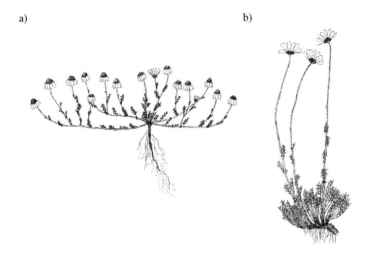

Fig.1.2. General habit of a) *A. secundiramea* (annual) and b) *A. cretica* subsp. *columnae* (perennial). (From Oberprieler, 1998).

Anthemis L. (Compositae, Anthemideae), with its c. 175 species (Oberprieler & al., 2006; see Appendix A1), is one of the largest genera within the tribe *Anthemideae* of the sunflower family (Asteraceae). It comprises annual or perennial herbs (Fig. 1.2) with imbricate involucral bracts, terete achenes and receptacular scales and it is characterized by a circum-Mediterranean distribution, with a geographical range that encompasses most of western Eurasia, North Africa, the Mediterranean region and a small part of eastern Africa. While only few species inhabit central Europe, the main centre of diversity can be found in south-west Asia, where more than 150 species, belonging to all of the presently accepted sections, occur (Oberprieler, 2001).

The taxonomic history of the genus is rather complex (see Appendix A2). A great variability at micro-morphological level associated with the difficulty of recognizing unique morphological characters that discriminate different taxa have resulted in the description of a large number of sections, series, species and subspecies. In its traditional circumscription *Anthemis* included the subgenus *Cota* J. Gay (i.e. Wagenitz, 1968; Bremer & Humphries, 1993) that today, mainly on carpological grounds, is recognized as a separate genus (Oberprieler & al., 2006). While representatives of *Anthemis* are characterized by actinomorphic achenes, round or tetragonal in cross section, *Cota* exhibits disymmetrical, dorso-ventrally flattened achenes (Fig. 1.3). Moreover, the epicarpic cells of *Anthemis* are filled with sand of calcium-oxalate crystals, while in *Cota* they either lack or contain single, large crystals (Oberprieler, 1998).

a)   b)

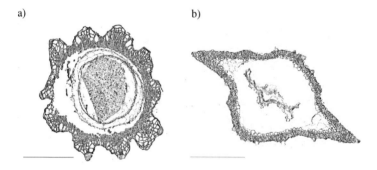

Fig. 1.3. Cross section of achenes of disc florets of a) *Anthemis zaianica* and b) *Cota austriaca*. Scale bar = 0.3 mm. (From Oberprieler, 1998).

Within the genus *Anthemis* five sections are recognized. Representatives of *Anthemis* sect. *Hiorthia* (DC.) R. Fern., which includes all polyploid species of the genus, are characterized by perennial habit and were often considered the most basal group of the genus (Meusel & Jäger, 1992). This section, described by Fernandes (1975) for Europe, corresponds to ser. *Rumata* Fed. (Fedorov, 1961), which includes other perennial species endemic to Caucasus (*A. marschalliana* and *A. fruticulosa*) or with a west-asian distribution (*A. calcarea*).

The total number of species included in this section is still not clear as there are many perennial species distributed throughout the whole range of the genus that were formally never included in any section.

All other sections include annual representatives. *Anthemis* sect. *Maruta* (Cass.) Griseb. consists of 14 annual species mainly distributed in the Middle East (Yavin, 1970) and its members are characterized by subulate receptacular pales and terete achenes. Fedorov (1961) described three series, namely ser. *Cotulae* Fed., ser. *Microcephalae* Fed. and ser. *Odontostephanae* Fed., each represented by two species (Fedorov, 1961). The relationships between ser. *Odontostephanae* (which includes the two closely related species *A. odontostephana* and *A. tubicina*) and the section *Maruta* were already assumed by Boissier (1875), but were then rejected by Eig (1938) due to the marked differences to the other representatives of the section in the general habit of involucral bracts, flowers and achenes, which are distinctly dilated at the base. When Yavin (1970) revised this section, she proposed the sectional or even generic rank for ser. *Odontostephanae*. She also included in the ser. *Cotulae* other 10 species with a Mediterranean distribution, reaching the actual number of 14 species for section *Maruta*.

Until the revision of Yavin (1972), all other species belonging to the genus *Anthemis* were included in *A.* sect. *Anthemis*, sharing the annual habit as the only synapomorhy. This section included species from all over the total geographical range of *Anthemis* and was the most species-rich section of the genus. Fedorov (1961), referring only to the species distributed in the former USSR, had identified only two series (ser. *Arvenses* Fed. and ser. *Candidissimae* Fed.). In a more comprehensive study, including all species of section *Anthemis* and based on the morphology of involucral bracts and achenes, Yavin (1972) added 14 new series and created two new sections: *A.* sect. *Chia* Yavin and *A.* sect. *Rascheyanae* Yavin. Section *Chia* includes only *A. chia*, an annual species with a mediterranean distribution that shows several unique morphological characteristics (acute involucral bracts with a broad, scarious rust-coloured margin; receptacular bracts falling off readily at maturity; dark, deeply grooved achenes).

Section *Rascheyanae* includes c. 15 annual species with the easternmost distribution of the whole genus (from Palestine to Syria, Iraq, Iran, Turkmenistan, Afghanistan, Pakistan). Its species are characterized by achenes 4-5 times longer than broad, an exceptional length to width ratio compared to the other species belonging to the genus *Anthemis*, where achenes are only 2-3 times longer than broad (Yavin, 1972). Eig (1938), who had previously defined the 'group of *A. rascehyana*', assigned four species to this group. Later, Yavin (1972) formally described the section *Rascheyanae* Yavin and added other seven species. The number of species was finally increased by Iranshahr (1982a, b), who described four new annual species from Iran, Iraq and Afghanistan (*A. freitagii, A. hamrinensis, A. gracilis* and *A. rhodocentra*).

The genus *Cota*, which comprises c. 35 species mainly distributed in the eastern part of the Mediterranean or in the Near East, was divided into two sections (Fernandes, 1975) that are distinguished on the basis of their habit. *Cota* sect. *Cota* includes all annual representatives of the genus, while *C.* sect. *Anthemaria* Dum. comprises the perennial species. In a former, partial classification for the former USSR however (Fedorov, 1961) five series were identified on the basis of habit and a combination of four morphological characters (involucral bracts, flowers, stems and leaves).

Oberprieler (1998) revised *Anthemis* in western and central North Africa and, later, carried out a preliminary investigation of the phylogenetic relationships among the sections based on the nrDNA ITS sequences of 27 representatives of *Anthemis* and *Cota* (Oberprieler, 2001). Even though the species sampling was poor, problems concerning the infrageneric classification were already observed, as the monophyly of almost all sections was not supported.

Thesis outline

To study both the patterns of plant biodiversity and the processes leading to diversity and endemism, two fundamentally different approaches are usually utilised: In a macroevolutionary, top-down approach phylogenetic reconstructions for a preferably completely sampled group of organisms are used as a backbone to reconstruct its biogeographical history, while in a microevolutionary, bottom-up approach, population genetic methods are applied for the reconstruction of spatial and temporal diversification patterns of single or closely related species (phylogeography; Avise, 2000) and for the study of processes connected with the adaptation of populations to certain environmental factors.

Recent advances in molecular phylogenetics have provided powerful tools for evolutionary analyses both at a macro- and a microevolutionary scale. DNA sequence data provide a relative wealth of characters for phylogenetic reconstructions at the species level. The temporal dimension of the diversification of lineages can be analysed through newly developed methods that, relaxing the strict molecular clock assumption, have become much more realistic (Linder, 2006). Notably, the relaxed Bayesian molecular clock (BEAST; Drummond & Rambaut, 2007) is one of the most sophisticated methods available as it estimates phylogeny and divergence dates simultaneously. Moreover, robust statistical analytical methods were developed to reconstruct the evolution of both discrete (such as geographical areas; e.g., DIVA; Ronquist, 1996) and continuous (such as climatic variables; e.g., 'MaxMin Coding'; Hardy & Linder, 2005) characters in a given phylogeny, thus allowing the identification of the relative importance of geographical vs. climatological differentiation processes in the history of a genus. At microevolutionary scale, DNA haplotypes and/or DNA fingerprinting shed light on population-level interactions, both in space and through time (phylogeography; Avise, 2000). Recent developed methods of eco-climatological niche modelling (e.g. Maxent; Phillips & al., 2006) allow investigating the role of climatological forces also on a local scale.

This thesis utilises both macro- and microevolutionary approaches to gain further understanding of evolutionary processes in the Mediterranean biome. While the genus *Anthemis* L. serves as a typical and species-rich representative of the Mediterranean and circum-Mediterranean vegetation for general biogeographical reconstructions (Chapter 2 and 3), the central Mediterranean *A. secundiramea* group acts as a study group for evolutionary processes in a Mediterranean archipelago with continental and volcanic islands (Chapter 4).

In the first part (Chapter 2) the monophyly of the two closely related genera *Anthemis* and *Cota* and the relationships among the infrageneric groups are examined on the basis of a comprehensive sampling including ca. 75% of all known species belonging to the two genera. A dated molecular phylogeny is based on sequence information from two plastid regions (*psbA-trnH* and *trnC-petN* spacer regions of the chloroplast DNA) and one nuclear marker (internal transcribed spacer (ITS) of the nuclear ribosomal DNA (nrDNA)) and is supported by 25 micro-morphological features.

The goal of the second part (Chapter 3) is to contribute to an assessment of the relative importance of geographical vs. climatological forces involved in the diversification of the genus. To achieve this, the strategy is to quantify the connected differentiation processes in terms of geographical vicariance events (as a measure of geographical mobility) and eco-climatological differentiation events ('climatic vicariance'; as a measure of ecological flexibility on niche evolution) in the evolution of the genus *Anthemis*.

The third part (Chapter 4) aims to identify the role of geographical vs. climatological differentiation processes at a local scale and in a time-window limited to the last 5 Ma. The study group (*A. secundiramea* group) is made up of six species distributed across the Sicilian Channel (N Africa, Sicily and its surrounding islands and islets). The analyses focus on the effects of some recent geographical and climatological events, such as those related to the glaciations cycles of the Quaternary, to genetic variation recorded in a mainland system as compared to the one found in a system of islands.

Chapter 2

Taxonomy and temporal diversification
of the genus A*nthemis* s.l. (Compositae, Anthemideae)
based on micromorphological characters
and three molecular markers

Materials and methods

Taxon sampling and sequencing

A comprehensive sampling including c. 75% of all known species belonging to the closely related genera *Anthemis* and *Cota* was carried out, and 141 species (168 accessions, see Appendix A1) were sampled from herbarium specimens held in B, W, WU and S. The taxon sampling included representatives of all currently recognized sections and comprised the full geographical range and all major habitats of the two genera *Anthemis* and *Cota*. To test the monophyly of infrageneric taxa, 19 representatives of the tribe Anthemideae were used as outgroups (Oberprieler & al., 2007). After performing preliminary analyses with different chloroplast DNA (cpDNA) markers, the two noncoding regions *psbA-trnH* and *trnC-petN* were chosen on the basis of good PCR amplification and sequence variability. The sequences generated from these markers were combined with those already available from former published accessions (Oberprieler, 2001; Oberprieler & Vogt, 2006; Lo Presti & Oberprieler, 2009; for details, see chapter 3 and Appendix A1) for the internal transcribed spacer (ITS) region of the nuclear DNA (nrDNA).

DNA was extracted from 20-30 mg crushed leaf material according to a modified protocol based on the method of Doyle & Doyle (1987). The chloroplast *psbA-trnH* spacer was amplified using the primers *psbA* and *trnH-*

(GUG) (Hamilton, 1999), while for the amplification of the *trnC-petN* region the primers *trnC* (Demesure & al., 1995) and *petN1R* or *petN2R* (Lee and Wen, 2004) were used. Amplification reactions followed Oberprieler & al. (2007). The polymerase chain reaction (PCR) products were purified with Agencourt AMPure magnetic beads (Agencourt Bioscience Corporation, Beverly, Massachusetts, USA) and cycle sequencing reactions were performed using the DTCS Sequencing kit (Beckman Coulter, Fullerton, California, USA), following the manufacturer's manual. The fragments were separated on a CEQ8000 sequencer (Beckman Coulter, Fullerton, California, USA). The DNA sequences obtained were carefully checked for presence/absence of polymorphic sites. We used the IUPAC (International Union of Pure and Applied Chemistry) ambiguity codes to indicate single nucleotide polymorphisms. A site was considered polymorphic when more than one peak was present on the electropherogram, with a minimal intensity of 25 % for the weaker peak compared to the stronger signal (Fuertes Aguilar & al., 1999; Mansion & al., 2005). The observation that these polymorphisms were very rare in the directly sequenced PCR products (2.05% and 1.34% of the total number of sequenced position for *psbA* and *trnC*, respectively) allowed us to abstain from cloning of products. Sequences obtained were aligned separately using Clustal W (Thompson & al., 1994) as implemented in BioEdit version 7.05.2 (Hall, 1999) and each alignment was optimized manually. An equivocal region in the alignment of the *trnC-petN* spacer consisting of 52 bp was excluded from all subsequent analyses. Gaps were coded as binary characters using the simple gap coding method of Simmons and Ochoterena (2000) implemented in GapCoder (Young and Healy, 2003).

Phylogenetic analyses

Each of the three partitions (*psbA-trnH*, *trnC-petN*, ITS), consisting of a stretch of nucleotides and the associated gaps, was at first analysed independently. The maximum parsimony (MP) analysis was performed using TNT version 1.1 (Goloboff & al., 2008), a software that implements several algorithms, such as ratchet (Nixon, 1999), tree-drifting, tree-fusing and sectorial searches (Goloboff,

1999), much more effective than simple branch-swapping, especially in large data sets (Goloboff & al., 2008). Following the suggestions of Drummond & al. (2007), we conducted a first search with 10 random addition sequences, each with tree bisection–reconnection (TBR) branch swapping. We then adopted the length-score found in the first search as starting point for the second search, made with the following parameters of the new technology search: 100 sectors of size 45 to 100 for the random sectorial searches (RSS), 40 and 30 rounds in tree-drifting and ratchet, respectively, and 10 initial replications to submit to the fusing algorithm. Character states were treated as unordered and unweighted. The search was conducted until the consensus became stable. To test the robustness of individual branches, bootstrap (BS) values (Felsenstein, 1985) were calculated. The bootstrap analysis was performed using 1,000 replicates, random addition sequence, TBR branch swapping and saving 1 tree per replicate. The search parameters were the same as adopted for the tree search.

A Bayesian inference (BI) phylogenetic analysis was performed with MrBayes version 3.1.2 (Ronquist & Huelsenbeck, 2003), implementing a GTR model with an invariate gamma distribution ('Lset NSt=6 rates=invgamma'), without fixing rates and nucleotide frequencies, as these parameters are estimated from the data during the analysis. Substitution models and rates of substitution were allowed to vary among the parameters ('unlink' command and 'ratepr=variable') and a binary model ('Lset coding=variable') was applied to the coded gaps (Ronquist & al., 2005). The analyses were conducted using up to five heated chains and one cold chain, with a chain heating parameter value of 0.08 - 0.2. The MCMC chains were run for up to 10,000,000 generations, with trees sampled every 1,000th generation. Four independent runs were conducted for each partition. Longer runs and greater number of cold chains as suggested by default settings were required to achieve convergence (Bergh and Linder, 2009), which was checked by examining the average standard deviation of split frequencies and by comparing likelihood values and parameter estimates from different runs in Tracer v. 1.3 (Rambaut and Drummond, 2003). A 50% majority rule consensus tree of the trees sampled after the achieving of convergence was computed.

To test for incongruence in the different partitions, we compared trees resulting from the three independent analyses (*psbA-trnH*, *trnC-petN* and ITS) and we searched for well-supported conflict among them, defined as bootstrap value ≥ 75% or PP value ≥ 0.95 (Wiens, 1998; Eldenäs and Linder, 2000; Bergh and Linder, 2009). Owing to the lack of conflicts among the two cpDNA data partitions, we combined them after excluding all taxa with missing data in one of the two partitions.

Divergence time estimation

Divergence times were estimated in a Bayesian framework by using the software BEAST version 1.4.7 (Drummond & Rambaut, 2007). BEAST has the advantage to test evolutionary hypotheses without constraining to a pre-defined tree topology as phylogeny and divergence dates are simultaneously estimated under a relaxed molecular clock (Drummond & al., 2006). After excluding all taxa with missing data and/or showing well-supported incongruence among the different partitions, we performed the analysis on all partitions together (*psbA-trnH*, *trnC-petN*, ITS). We have desisted from a comprehensive analysis with all taxa, as it was demonstrated that the combined analysis of incongruent data partitions introduces a bias in the divergence time estimates (Pfeil, 2009). In the reduced dataset, including 116 accessions, 6 representatives of the subfamily Asteroideae were added (members of the tribes Senecioneae, Astereae, Calenduleae, Gnaphalieae and Heliantheae s.l.). We used the software BEAUti version 1.4.7 (Drummond & Rambaut, 2007) to create the XML input file for BEAST. We defined the model of evolution (GTR+I+Γ), a relaxed molecular clock that assumes independent rates on different branches ('relaxed clock: uncorrelated lognormal'), and we also set a Yule prior that assumes a constant speciation rate per lineage, as suggested for species-level phylogenies (Drummond & al., 2007). The 'mean.Rate' parameter had a uniform prior between 0.0 and 0.1, the 'coefficientOfVariation' and the 'covariance' prior parameters were uniform between (0.0; 1.0) and (-1.0; 1.0) respectively. The efficiency of the MCMC chain was improved by altering the weight of the operators that work on the 'treeModel', as suggested by Drummond & al. (2007). The XML file created with BEAUti was then edited manually to allow substitution models and rates of substitution to be estimated independently among the data partitions. The runs were started on random trees. After conducting several short runs (up to 10,000 generations) to optimize the parameters as proposed by the performance suggestions in the output, two runs of 10,000,000 generations, sampled every 1,000th generation, were performed. To determine convergence, the results of the two runs were analysed using Tracer version 1.4 (Drummond & Rambaut, 2007). Of the total 20,000 trees, 16,000 trees were sampled, combined with LogCombiner version 1.4.8 (Drummond & Rambaut, 2007) and finally summarized in a maximum clade credibility tree (which is the tree with the highest product of all the clade posterior probabilities) with a posterior probability limit set to 0.5 in TreeAnnotator version 1.4.8 (Drummond & Rambaut, 2007).

To calibrate the tree, we defined taxon subsets that were not constrained to be monophyletic and we set prior distributions on the corresponding divergence times. The root node represents the basal node of the subfamily Asteroideae. The age of this node was calibrated on the basis of two previous molecular dating studies, both conducted with external calibration (Kim & al., 2005; Hershkovitz & al., 2006). Hershkovitz & al. (2006) used a date of 128 Ma for the crown Asteridae (Bremer & al., 2004), while Kim & al. (2005) adopted both non parametric rate smoothing with an internal calibration point and fixed substitution rates from other angiosperm families. The three different methods converge well in the results, as the crown age of the Asteroideae is c. 30-29 Ma (Hershkovitz & al., 2006), 29-26 Ma (Kim & al., 2005; non parametric rate smoothing or 39-35 Ma (Kim & al., 2005; fixed substitution rate). Therefore, we set a normal prior distribution with a mean of 31.3 Ma (the mean of all values) and a standard deviation of 3.1 Ma. The second calibration point used is the root node of the tribe Heliantheae s.l. (here represented by the two species *Tagetes patula* and *Helenium autumnale*). The first fossil of the *Ambrosia*-type pollen that characterizes this tribe was dated to be 25-35 Ma old (Graham & al., 1996) and we used a normal prior distribution with a mean of 30.0 Ma and 95% confidence interval that covered the entire interval of uncertainty (25.1 – 34.9 Ma). The split between the Mediterranean representatives of *Anthemis/Cota* and the monophyletic group of *Gonospermum canariense* + *Lugoa revoluta*, endemic to the Canary Islands Tenerife, Gomera, La Palma and El Hierro was adopted as third calibration node. The uncertainty for this node was accommodated by specifying a normal distribution with a mean of 10.4 Ma and 95% confidence interval comprised between the earliest and the latest estimated age of Tenerife (11.9-8.9 Ma; Guillou & al, 2004). The last taxon subset included *A. aetnensis* Spreng. and its sister species *A. cretica* L. subsp. *cretica*. *Anthemis aetnensis* is endemic to Mt. Etna, the largest active volcano in Europe (Rittmann, 1973; Frazzetta and Villari, 1981; Marty & al., 1994), and it is one of the few species that reaches an altitude of 3050 m, which is the limit of the vegetation on this volcano (Poli, 1965). We considered the age of Mt. Etna to calibrate the split between the two sister taxa and we accommodated the uncertainty associated with the colonization event by specifying a normal prior distribution with a 95% confidence interval ranging between 300 ka (the upper age limit of Mt. Etna; Rittmann, 1973; Duncan & al., 1984; Chester & al., 1985) and 100 ka (mean = 200 ka).

To check for robustness of dating information obtained through BEAST, divergence times were also estimated in r8s version 1.70 (Sanderson, 2003). A comprehensive Bayesian analysis including all partitions (*psbA-trnH, trnC-petN,* ITS) was carried out on the same reduced dataset with 116 accessions adopted for BEAST and with the same parameters used for each separate partition. A penalized likelihood analysis (PL; Sanderson, 2002) was performed on one of the Bayesian trees obtained. The settings of the analysis are summarized in Table 2.1. The optimal smoothing level, found to be 0.005 and chosen via the cross-validation procedure (which gives an indication of the clock- vs. non-clocklike behaviour of the data; Sanderson, 2002), was then used in the estimation of divergence times in the phylogram obtained from the BI analysis and in the computation of the associated error rate. The estimates of the error rate were obtained by generating 1,000 phylograms in PAUP* (Swofford, 2002) from bootstrap resampling of the original dataset. Nodal ages obtained from all trees were finally summarized in r8s (using the 'profile' command) and the resulting standard deviations were used to calculate the 95% confidence intervals for the estimated ages of the original phylogram.

For the major clades, the logarithm of the cumulative number of lineages was plotted against the absolute date of each node obtained through BEAST (lineages-through-time plot, LTT). This graph, which assumes a constant extinction rate, gives a coarse impression of the temporal diversification of lineages (Barraclough & Reeves, 2005; Kadereit & Comes, 2005; Savolainen & Forest, 2005).

Micro-morphological analyses

For a subset of taxa, including 43 representatives of all currently recognized sections of the two genera *Anthemis* and *Cota* and the outgroup species *Tripleurospermum parviflorum*, *Tripleurospermum inodorum* and *Nananthea perpusilla*, 25 morphological characters were scored, in order to test whether the groups obtained from the molecular phylogeny were also supported on morphological grounds. For the infrageneric classification of the genera *Anthemis* and *Cota*, the micro-morphological characters (in particular achenes and pales) are essential (see Fernandes, 1976; Oberprieler, 1998, 2001). For this reason, we have focused on the 25 micro-morphological characters summarized in Table 2.1. Twenty of them were treated as categorical and up to 4 categories per characters were defined, while the other five were considered continuous. Disc flowers, achenes and leaves were boiled in distilled water, mounted under a coverglass in Kaiser's glycerin gelatine (Merck, Darmstadt, Germany) and then classified/measured under a stereo-light microscope (Axioskop 2 Mat Zeiss, Göttingen, Germany). The objects were finally documented with an AxioCam HRc camera (Zeiss, Göttingen, Germany) and analysed with the software Axiovision (AxioVs40 vs. 4.6.1.0 Zeiss, Göttingen, Germany). A Principal Component Analysis (PCA) was performed based on all 25 characters in order to obtain an interpretable two-dimensional ordination of taxa according to their morphological similarities.

Table 2.1. Micro-morphological characters adopted for the morphological analyses of 46 taxa, along with factor loadings of the PCA (see Fig. 2.6).

Abb.	Part of the plant	Character Name	Character Specification	PCA Analysis	
				PC1 Loadings	PC2 Loadings
M1	Leaf	Hair type on leaves.	0: no hairs 1: basifixed hairs 2: equally medifixed hairs 3: irregularly medifixed hairs	0.205	-0.261
M2	Leaf	Resin ducts along the midveins of the pinnae.	0: present 1: absent	-0.499	-0.249
M3	Leaf	Shape of the terminal part of the pinnae.	0: rounded 1: shortly cuspidate 2: cuspidate 3: long cuspidate	0.341	0.308
M4	Flower-corolla	Arrangment of glands in the upper part of the corolla.	0: no glands 1: few glands 2: glands only along the vascular bundles 3: many glands everywhere	0.182	0.000
M5	Flower-corolla	Average number of xylem elements forming the corolla vascular bundle.	-	-0.303	0.445
M6	Flower-corolla	Length of a corolla vascular bundle relative to the length of the corolla tube (excluding corolla lobes).	-	-0.005	0.372
M7	Flower-corolla	Resin ducts in the vascular bundles of the corolla.	0: present 1: absent	-0.096	-0.001
M8	Flower-corolla	Resin sacs in the corolla lobes.	0: present 1: absent	-0.538	0.023
M9	Flower-corolla	Shape of the corolla lobes.	1: narrow 2: wide	-0.549	-0.096
M10	Flower-corolla	Shape of the cucullate appendages of the corolla lobe.	1: acute 2: rounded	-0.419	-0.120
M11	Flower-corolla	Dimension of the cucullate appendages of the corolla lobe.	1: small 2: large	0.622	-0.033
M12	Flower-corolla	Inflation of the corolla base at maturity.	0: not inflated 1: inflated only in the basal part 2: inflated up to the middle	0.284	-0.127

Table 2.1. Continued.

Abb.	Part of the plant	Character Name	Character Specification	PCA Analysis	
				PC1 Loadings	PC2 Loadings
M13	Flower-stamen	Constriction between the anther and the apical appendages of the stamen.	0: present 1: absent	0.867	-0.007
M14	Flower-stamen	Shape of the anthers-appendages.	1: triangular 2: ovate 3: elliptic 4: rounded	0.072	-0.207
M15	Flower-stamen	Ratio between the width of the appendages at their broadest point and the total length of the appendages.	-	0.888	-0.057
M16	Flower-stamen	Shape of the filament collar.	1: inflated 2: bottle shaped 3: triangular 4: lineal	0.120	-0.019
M17	Flower-stamen	Width of the filament collar (expressed in cell raws).	-	-0.179	0.454
M18	Flower-stamen	Length of the filament collar (expressed in cell numbers).	-	-0.192	0.414
M19	Flower-stigma	Resin ducts in the stigmatic branches.	0: present 1: absent	0.225	-0.006
M20	Achene	Calcium-oxalate in the epicarpic cells.	0: sand of calcium-oxalate 1: single, large crystals of calcium-oxalate	0.139	0.810
M21	Achene	Dimension of the mucilage cells in the epicarp.	0: small 1: large and protruding	-0.177	-0.434
M22	Achene	Glandular hairs on epicarp.	0: present 1: absent	0.166	-0.406
M23	Achene	Number of ribs of achenes.	1: less than 10 ribs 2: 10 ribs 3: more than 10 ribs	0.162	0.296
M24	Achene	Shape of the achene in cross section.	1: dorso-ventrally flattened 2: rounded or tetragonal	0.204	0.756
M25	Achene	Shape of the wall of the testa-cells.	0: straight 1: undulated	0.266	-0.605

Results

Sequence variation and phylogenetic analyses

The alignments of the two cpDNA regions are quite different in their characteristics. The *psba-trnH* spacer is shorter than the *trnC-petN* region (485 bp for *psbA-trnH* vs. 812 bp for *trnC-petN*, respectively) but contains almost twice as much parsimony informative (PI) characters (24.2% for *psbA-trnH* vs 12.4% for *trnC-petN*). However, although the *trnC-petN* region shows low variation, there is little homoplasy (homoplasy index, HI=0.19) and the phylogenetic signal is therefore rather consistent. In both alignments the larger amount of PI characters are derived from indel data, with almost half of them (53.8% for *psbA-trnH* and 47.1% for *trnC-petN*, respectively) being PI. No well-supported incongruence (BP ≥ 75% and PP ≥ 0.95) between the two plastid partitions (*psbA-trnH* and *trnC-petN*) was detected and other events of incongruence supported with lower BP or PP were considered to be insignificant (Hillis and Bull, 1993). The total length of the combined cpDNA dataset including the indels was 1,439 bp, with 228 bp (15.8%) of PI characters and a rather law homoplasy level (HI=0.27). The length of the ITS alignment including ITS1 (241-264 bp) and ITS2 (198-224 bp) was considerably shorter than the cpDNA dataset (660 bp). It provided about twice the amount of variation (43.6%), but at the same time about twice the homoplasy (HI=0.58). The variation was mainly due to point mutations, equally distributed between substitutions (43.1%) and deletions (45.5%). Character statistics for all datasets are summarized in Table 2.2.

The parsimony analysis of the combined cpDNA dataset yielded 192 equally most-parsimonious trees with a length of 710 steps, a consistency index (CI) of 0.73 and a retention index (RI) of 0.89 (autoapomorphies excluded; tree not shown). The analysis of the ITS dataset yielded 109 equally most parsimonious trees (1,486 steps, CI=0.42, RI=0.77; autoapomorphies excluded; tree not shown). In the cpDNA dataset, the number of resolved nodes in both MP and BI analysis is not very high, reaching only 42.7% of the maximum possible theoretical number of internal branches (calculated as number of taxa included in the analysis – 2; see Bergh & Linder, 2009). However, in the strict consensus tree of the MP analysis (not shown), 50% of the resolved nodes are supported by a bootstrap value $\geq 75\%$ and in the 50% majority-rule consensus tree of the partitioned Bayesian analysis (see Fig. 2.2) the percent for the nodes supported with a $PP \geq 0.95$ reaches 64.1%. In the ITS dataset a different situation is observed. In both the MP (not shown) and the BI trees (see Fig. 2.1) up to 64.8% of the maximum possible number of nodes are resolved, but only 26.9% (in the MP tree) and 53.4% (in the BI tree) of them are well supported. This fact is not surprising, as the cpDNA dataset includes less PI characters than the ITS dataset, but also less homoplasy (HI=0.27 and HI=0.58 for cpDNA and ITS, respectively).

Table 2.2. Summary statistics of the ITS, cpDNA and combined datasets adopted in the analyses.

	psbA-trnH	*trnC-petN*	Combined plastid	ITS
Total number of taxa (Number of outgroup taxa)	166 (14)	164 (13)	152 (13)	184 (19)
Number of characters	485	812	1297	526
Number of Parsimony Informative characters (%)	90 (18.5)	83 (10.2)	166 (12.8)	227 (43.1)
Number of indels	93	51	142	134
Number of Parsimony Informative indels (%)	50 (53.8)	24 (47.1)	62 (43.7)	61 (45.5)
Total number of characters	578	863	1439	660
Total Number of Parsimony Informative characters (%)	140 (24.2)	107 (12.4)	228 (15.8)	288 (43.6)
Number of steps (MP)	383	306	710	1486
CI (Consistency Index)	0.702	0.814	0.73	0.424
RI (Retention Index)	0.872	0.93	0.887	0.768
HI (Homoplasy Index)	0.298	0.186	0.27	0.576
N nodes (%) resolved in MP strict consensus	74 (45.1)	42 (25.9)	52 (34.6)	78 (42.8)
N nodes (%) with $BS \geq 75$	21 (12.8)	15 (9.3)	26 (17.3)	21 (39.6)
N nodes (%) resolved in MB maj-rule consensus	54 (32.9)	50 (30.9)	64 (42.7)	118 (64.8)
N nodes (%) with $PP \geq 0.95$	31 (18.9)	30 (18.5)	41 (28.3)	63 (34.6)

Next two pages: Fig. 2.1. BI tree of nrDNA ITS data for 184 taxa. The legend (only in Fig. 2.1a) shows the traditional sectional subdivision for the two genera *Anthemis* and *Cota*. Numbers above the branches indicate support values (only PP values ≥ 0.95 are shown). Taxa that show well-supported conflict (PP ≥ 0.95 or BP from MP analysis $\geq 75\%$) among ITS and cpDNA datasets are underlined. The symbol "#" followed by a shortcut for the name of the taxon identifies the 46 taxa morphologically analysed. (a): upper part of the tree. (b): lower part of the tree.

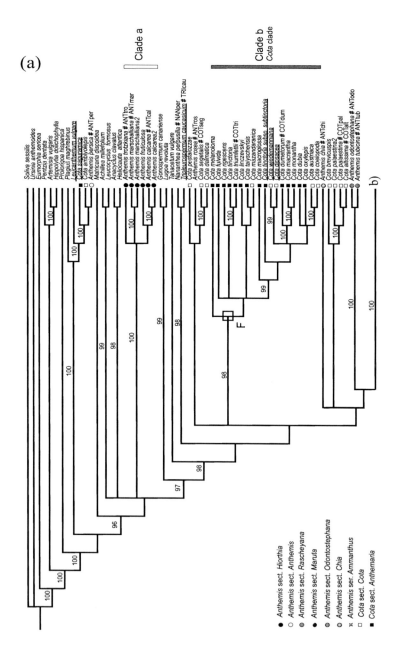

- ● *Anthemis* sect. *Hiorthia*
- ○ *Anthemis* sect. *Anthemis*
- ◓ *Anthemis* sect. *Rascheyana*
- ● *Anthemis* sect. *Maruta*
- ◕ *Anthemis* sect. *Odontostephana*
- ◔ *Anthemis* sect. *Chia*
- ⋈ *Anthemis* ser. *Ammanthus*
- □ *Cota* sect. *Cota*
- ■ *Cota* sect. *Anthemaria*

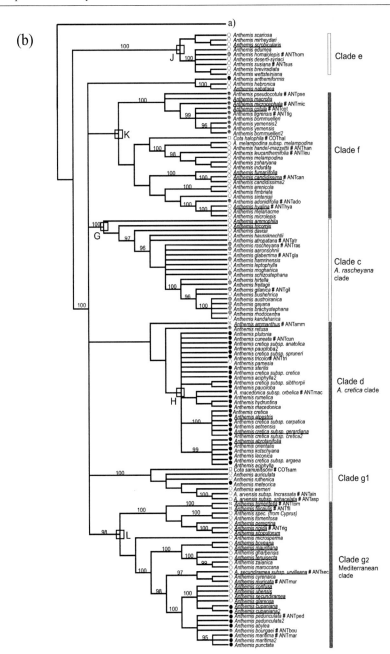

Topology

Both cpDNA and ITS datasets do not support the monophyly of the genus *Anthemis*. There are in particular four species (the perennials *A. trotzkiana*, *A. calcarea*, *A. marschalliana* and *A. fruticulosa*, endemic to the Caucasian region) that are more basal than representatives of the genus *Cota* or the outgroup taxa *Tanacetum* and *Gonospermum* and that form a well supported clade in the BI analysis of the ITS dataset (Fig. 2.1). The genus *Cota*, with the exception of some taxa (8 in the ITS-BI tree, 7 in the cpDNA-BI tree), is monophyletic both in the ITS and in the cpDNA BI trees (*clade b*; Figs. 2.1, 2.2) but the relationship with the main core of *Anthemis* remains unclear, as the basal nodes of both trees are not supported. It is interesting to note that the annual Mediterranean species *A. filicaulis*, *A. tomentella*, *A. rigida*, *A. scopulorum* and *A. ammanthus* are nested within *Cota* only in the cpDNA-BI tree (*clade b*, Fig. 2.2a), while in the ITS- BI tree they are included within *Anthemis* (*clade g2*, Fig. 2.1). Within the *Anthemis* main clade, several subclades can be recognized: *clade c* (the *A. rascheyana* clade) includes annual species widespread in western Asia, while *clade d* (the *A. cretica* clade) comprises almost all perennial species of the genus *Anthemis*, being distributed in the mountains of the circum-Mediterranean region (Figs. 2.1, 2.2). *Clade g2*, which encompasses annual and perennial species widespread in the central and western Mediterranean area, is well supported in the ITS-BI tree (Fig. 2.1). In the cpDNA-BI tree, it represents a subset of a more comprehensive, not well-supported clade (*clade g*, Fig. 2.2b) that includes also species distributed in the eastern Mediterranean region, such as *A. retusa* or *A. tomentosa*. The remaining species with an eastern-most distribution are grouped together in one

---

Next two pages: Fig. 2.2. BI tree of cpDNA (*psbA-trnH* and *trnC-petN* spacer regions) data for 152 taxa. The legend (only in Fig. 2.2b) shows the traditional sectional subdivision for the two genera *Anthemis* and *Cota*. Numbers above the branches indicate support values (only PP values ≥ 0.95 are shown). Taxa that show well-supported conflict (PP ≥ 0.95 or BP from MP analysis ≥ 75%) among ITS and cpDNA datasets are underlined. The symbol "#" followed by a shortcut for the name of the taxon identifies the 46 taxa morphologically analysed.

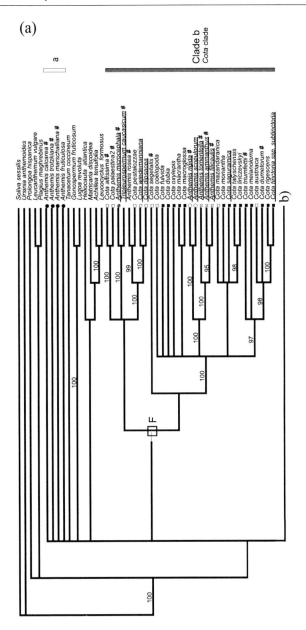

(a)

(b)

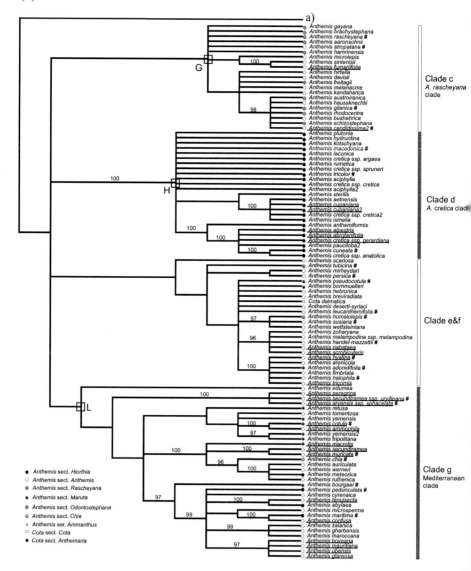

clade in the species with an eastern-most distribution are grouped together in one clade in the cpDNA-BI tree (*clade e&f*, Fig. 2.2b), while they are included in two separate clades (*clade e* and *clade f*, Fig. 2.1) in the ITS-BI tree. We found rather many taxa (36, corresponding to 19.1% of the total number of taxa included in the analyses) with a well-supported conflicting placement in trees from the plastid and the nuclear datasets (BP ≥ 75% or PP ≥ 0.95). However, all (but two) incongruences involve tip nodes and half of it do not affect the position of a taxon within a bigger clade (see for example the position of *A. boveana, A. mauritiana, A. ubensis* and *A. glareosa* in *clade g* or that of *C. tinctoria* subsp. *tinctoria* in *clade b*; Figs. 2.1, 2.2).

The traditional subdivision in sections and series seems to be not entirely supported by this phylogenetic analysis as representatives of different sections or series belonging to the two genera *Anthemis* and *Cota* are widespread throughout the trees. Obvious examples are the perennial *Anthemis* section *Hiorthia*, split at least in three parts, and *A.* section *Anthemis* (Figs. 2.1, 2.2).

Divergence time estimation of the combined nuclear and plastid dataset

The alignment of the combined dataset for 116 taxa was 1,865 bp long and the resulting BEAST maximum clade credibility (MCC) tree with the highest posterior probabilities is shown in Fig. 2.3. The topology of this tree is intermediate between the ITS-BI tree and the cpDNA-BI tree: for example, the annual species widespread in the eastern Mediterranean region that form two separate clades in the ITS analysis (*clade e* and *clade f*, Fig. 2.1) and are included in one large clade in the cpDNA tree (*clade e&f*, Fig. 2.2b), appear as two separate but sister clades in the BEAST tree (*clade e* and *clade f*, Fig. 2.3).

Next page: Fig. 2.3. Maximum-clade-credibility tree of the combined dataset (ITS, *psbA-trnH* and *trnC-petN*) for 116 taxa obtained from the BEAST analysis. Time scale is in millions of years (Ma). Numbers above the branches indicate support values (only PP values ≥ 0.95 are shown). Nodes adopted for the calibration (A, B, C, D) are identified by the symbol "■". Age spans for nodes identified by "□" are provided in Table 2.3.

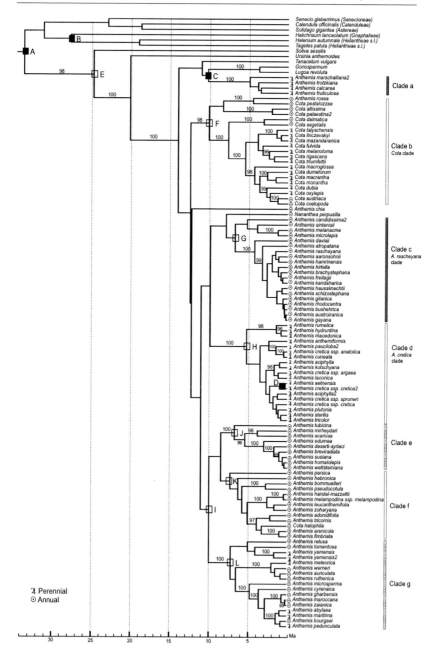

The mean number of substitutions per site per million years (mean.Rate) across the whole tree was estimated to be 0.0048 (0.0036-0.0062). All posterior distributions of calibration nodes match their priors quite good (Fig. 2.4), even though they show larger confidence intervals. The root node, which designates the crown age of the subfamily Asteroideae, is slightly shifted backwards in time relative to its prior and is estimated to be 33.2 ± 5.17 (Late Eocene-Early Oligocene). The derived crown age of the tribe Anthemideae is 24.5 ± 7.76 Ma (Late to Middle Oligocene). All age estimates generally agree with those obtained via a PL approach (Sanderson, 2002). However, we should notice some discrepancy in the mean age reconstruction of clades *b*, *c* and *g*, but their slightly older age may be related to the settings adopted for the age estimation: in contrast to BEAST, r8s implements only 'hard' upper and lower

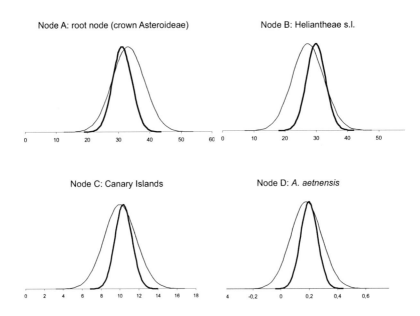

Fig. 2.4. Prior (thick line) and posterior (thin line) age distributions of the four calibration nodes in the BEAST analysis. Time scale is in Million of years.

cut-off bounds, and it is not possible to define a time range that incorporate more prior information about the timing of divergence (Yang & Rannala, 2006; Bergh & Linder, 2009). Our quite conservative settings (for node C, for example, we have defined a maximum age corresponding to the earliest estimated age of Tenerife; see Table 2.3) may, therefore, have resulted in older age estimates.

When disregarding some outliers (i.e. *A. chia*), we find that almost all major clades are not older than 10 Ma (Table 2.3), but exhibit different rates of lineage branching (Fig. 2.5). In particular, it is interesting to note that *clade g* shows a constant increase in the number of lineages in the last 10 Ma while *clade d* (*A. cretica* clade) and *clade c* (*A. rascheyana* clade) have radiated mainly in the last 2 Ma.

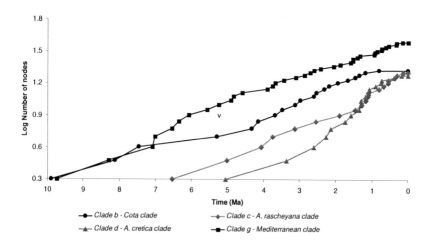

Fig. 2.5. Lineages-through-time (LTT) plot for clades *b*, *c*, *d*, *g* using dates obtained by the BEAST analysis.

Table 2.3. Summary of ages and age spans obtained from BEAST and r8s analyses.

Node (a)	Description	BEAST analysis (b)		r8s analysis (c)	
		Priors	Posteriors	Settings	Mean Age (± 95% CI)
A	Root (Crown Asteroideae)	31.30 (26.2-36.4)	33.21 (28.04-38.53)	31.30	31.30
B	Heliantheae s.l.	30.00 (25.1-34.9)	27.40 (22.30-32.37)	30.00	30.00
C	Canary Islands	10.40 (8.9-11.9)	10.02 (8.33-11.81)	maxage=11.9	11.90 (4.77-19.03)
D	*A. aetnensis*	0.20 (0.1-0.3)	0.18 (0.07-0.30)	maxage=0.20	0.20 (0.11-0.29)
E	Crown Anthemideae	-	24.5 (16.74-32.26)	-	28.21 (22.43-33.99)
F	Clade b (*Cota* clade) Clade c	-	9.89 (6.16-13.70)	-	14.38 (8.90-19.86)
G	(*A. rascheyana* clade) Clade d	-	6.53 (3.82-9.37)	-	11.96 (8.34-15.58)
H	(*A. cretica* clade)	-	5.05 (2.21-8.36)	-	7.12 (4.95-9.29)
I*	Mediterranean clade	-	9.74 (6.37-13.42)	-	-
J	Clade e	-	6.54 (3.50-9.95)	-	9.38 (6.82-11.94)
K**	Clade f	-	7.08	-	10.98
L	Clade g	-	7.01 (4.19-10.11)	-	11.02 (7.57-14.47)

(a) Node names as in Fig. 2.3.
(b) Values are in Ma and represent the mean age estimates and, in brackets, 95% confidence intervals.
(c) Values are in Ma and represent settings of the analysis, mean age estimates and, in brackets, 95% confidence intervals from 1,000 bootstrap replicates.
* Node absent in the MB tree adopted for the r8s analysis.
** Node too poorly supported both in BEAST and in r8s analyses, no confidence interval.

Micro-morphological analyses

The characters states obtained for all 46 taxa are summarized in Appendix A3. In the PCA of all 25 characters (Fig. 2.6; factor loadings in Table 2.1), the eight components with eigenvalues higher than 1 accounted 70% of the total variance. Since the first principal component of the PCA (PC1) explained only 14.7% of the total variance, a quite low correlation among these characters was revealed. The PC1 is strongly influenced by the characters of the apical appendage in the distal part of the stamen (M13 and M15 with loadings higher than 0.86; see Table 2.1) and moderately by the shape of the appendages of the corolla lobe (0.62).

Achenes characters (presence of calcium-oxalate crystals in the epicarpic cells and shape of the achenes in cross section) are positively correlated with each other and with the second principal component (PC2, accounting for 12.3% of total variation). While representatives of the genera *Cota* and *Anthemis* are well-separated on PC2, along the PC1 the three perennial species *A. marschalliana*, *A. trotzkiana* and *A. calcarea* cluster with the two taxa belonging to the section *Odontostephana* of the genus *Anthemis* (*A. odontostephana* and *A. tubicina*) and with three outgroup taxa (*T. perforatum, T. parvifolium* and *N. perpusilla*) on the left side of the plot and are well-isolated from the other representatives of the genus *Anthemis*.

Discussion

Congruence among chloroplast and nuclear datasets

The 152-taxa cpDNA dataset and the 184-taxa nrDNA dataset exhibit a moderate level of incongruence with 36 taxa showing well-supported conflicting placements in the resulting trees. Of the total 36 incongruences, 15 involve only tip nodes and do not affect the entire topology. We might attribute them to the method adopted: The two different datasets include different data, with the ITS dataset comprising more taxa (184) than the cpDNA dataset (152). Thus, the presence of taxa in only one of the two datasets may result in alternative relationships among the phylogenetically close taxa involved.

The conflicts among the remaining 21 taxa have to be reconducted to other causes, such as hybridization or incomplete lineage sorting (Funk & Omland, 2003; Wiens & al., 2003; Edwards & al., 2008; Edwards, 2009). Support for the occurrence of reticulate evolution was already provided by crossing experiments between *Anthemis, Cota* and *Tripleurospermum* (Mitsuoka & Ehrendorfer, 1972). Therefore, the possibility of occasional hybridization followed by selection of new fertile recombinants (Yavin, 1972; Grant, 1971; Stebbins, 1942) is plausible, even if the species involved are diploid. Moreover, as it is argued that *Anthemis* and *Cota* underwent a rapid radiation in the last 10-15 Ma (see below), incomplete lineage sorting may have played an important role, with speciation events occurring before sorting was complete (Funk & Omland, 2003).

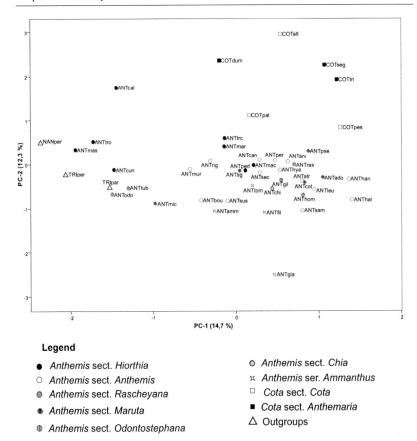

Fig. 2.6. Principal components analysis (PCA) of 25 morphological characters for 46 taxa belonging to the two genera *Anthemis* and *Cota* and to the two outgroups *Tripleurospermum* and *Nananthea*. The legend shows the traditional sectional subdivision for the two genera *Anthemis* and *Cota*. Shortcuts are given in Fig. 2.1.

Generic delimitation

The relative relationships between *Anthemis*, *Cota*, *Tripleurosperum* and *Nananthea* were already hypothesized to be very close using embryological (Harling, 1950, 1951, 1960), cytological (Uitz, 1970; Oberprieler, 1998) and chemical (Greger, 1977) data and in contrast to the placement in different subtribes by Bremer & Humphries (1993). Moreover, the observation of natural hybrids together with successful crossing experiments among taxa belonging to *Anthemis*, *Cota* and *Tripleurospermum* (Mitsuoka & Ehrendorfer, 1972) has shown the evolutionary affinities between these genera. In a preliminary molecular analysis, Oberprieler (2001) pointed out that the genus *Tripleurospermum* was consistently placed between *Anthemis* and *Cota* in MP, Maximum Likelihood and Neighbour-Joining trees. In the present analysis, we can confirm the affinities among these genera. *Tripleurospermum* and *Nananthea* form a well supported clade both in the ITS-BI tree (Fig. 2.1) and in the ITS-MP tree (results not shown), but their basal position in respect to both *Cota* and *Anthemis* is not supported.

Anthemis – The results obtained both by ITS and cpDNA analyses show that the genus *Anthemis* is not monophyletic. Four perennial species endemic to the Caucasian region and formally included in the section *Hiorthia* (*A. calcarea*, *A. marschalliana*, *A. trotzkiana* and *A. fruticulosa*) are more basal than the genus *Cota* or the outgroup taxa *Tanacetum* and *Gonospermum* both in the ITS-BI tree (*clade a*; Fig. 2.1) and in the cpDNA-BI tree (*clade a*; Fig. 2.2a). The micro-morphological analyses (Fig. 2.6) included three of these species (*A. calcarea*, *A. marschalliana* and *A. trotzkiana*) and show that they exhibit a unique set of micro-morphological characters: they are characterized by relatively acute and small corolla lobe appendages, by stamina with quite wide and long filament collars (c. 8-9 cells in width and up to 37 cells in length) and show no constriction between the anthers and their triangular apical appendage. The latter two characters are also found in the outgroup genera *Tripleurospermum* and *Nananthea* and can be therefore considered plesiomorphic. The acute shape of the corolla lobe appendages is also observed in other representatives of the genera *Anthemis* and *Cota* but it is the association of both small and acute corolla lobe appendages that is typical only for *A. calcarea*, *A. marschalliana* and *A. trotzkiana*. Moreover, eco-climatological analyses carried out on this group (Lo Presti & Oberprieler, 2009; see chapter 3) show that the four species are characterized by a montane climate, a state reconstructed to be plesiomorphic (as it is also shared by the common ancestors of the genera *Anthemis* and *Cota*).

Fedorov (1961) included *A. calcarea*, *A. fruticulosa* and *A. trotzkiana* in the ser. *Fruticulosae* Fed. and *A. marschalliana* in the ser. *Marschallianae* Fed. of sect. *Hiorthia*. The differences between the two series are based on floral characters (colour of the ligules, involucral bracts, size of the capitula), but they all share thick, woody rootstocks bearing many lignified based of last years' shoots. The suffrutescent habit typical for these species, that allows considering them not only as perennials but even as small subshrubs, let Fedorov (1961) to state that '[...] the most ancient species of the genus *Anthemis* have to be recognized [in] for example, *A. fruticulosa* and *A. trotzkiana* [...]' (Fedorov, 1961:10). The concordance of molecular, morphological and eco-climatological data supports the early divergence of these four species from the main clades of *Cota* and *Anthemis* and would argue for a treatment of these species at generic level.

In the ITS-BI tree, there are three other species that fall outside the main clades of *Anthemis* and *Cota* and that form a well supported clade with the genera *Leucanthemum* and *Plagius*: *A. persica* (an annual species endemic to Iran), *C. saguramica* (a perennial endemic to Caucasus) and *C. amblyolepis* (an annual species distributed in the east-Mediterranean region). While for the last one we lack plastid sequence information, it is noteworthy that, in the cpDNA-BI tree, *C. saguramica* and *A. persica* are placed within the well supported main clade of *Cota* and within *clade e&f*, respectively. The long terminal branches leading to the three taxa in the ITS-BI tree may argue that this position could be attributed to the effect of 'long branch attraction' (Hendy & Penny, 1989) rather than to an alleged hybridization with the west-Mediterranean genera *Leucanthemum* and/or *Plagius*, which belong to a different subtribe of Anthemideae.

Cota – The taxonomic position of *Cota* relative to *Anthemis* was discussed for a long time. *Cota* was alternatively treated as subgenus (Wagenitz, 1968; Fernandes, 1975, 1976; Bremer & Humphries, 1993) or even as section within *Anthemis* (i.e., Boissier, 1875; Fedorov, 1961; Grierson & Yavin, 1975). The formal recognition of *Cota* as independent genus by Greuter & al. (2003), who recover the genus name already introduced by J. Gay (in Gussone, 1845), is backed by phytochemical (Greger, 1969, 1970), caryological (Uitz, 1970), cytogentic (Mitsuoka & Ehrendorfer, 1972) and morphological evidence (Oberprieler, 1998, 2001). Previous molecular analyses (Oberprieler, 2001), based on one nuclear marker (nrDNA ITS) and including a subset of only 4 taxa belonging to *Cota*, had partially supported the monophyly of the genus. The present investigation, which comprises more than half of the formally accepted taxa included in the genus *Cota*, provides further evidence to the monophyly of the genus, both on molecular (Figs. 2.1, 2.2) and morphological basis. All investigated taxa (up to two, see below) share dorso-ventrally flattened achenes and single, large calcium-oxalate crystals in the epicarpic cells of achenes (Oberprieler, 1998, 2001, present study) so that they are distinctly separated from representatives of *Anthemis* (Fig. 2.6). The generic independence of *Cota* from *Anthemis* was already pointed out by the crossing experiments conducted by Mitsuoka & Ehrendorfer (1972), who have shown that the chromosomal affinity between representatives of *Cota* (*C. altissima*) and *Anthemis* (*A. cotula*) is even smaller than that obtained between *Anthemis* (*A. cotula*) and *Tripleurospermum* (*T. tetragonospermum*).

Five species formally included in *Anthemis* (*A. linczevskyi*, *A. mazandaranica*, *A. talyschensis*, *A. brevicuspis* and *A. macroglossa*) share the flattened achenes and the calcium-oxalate crystals with the other representatives of *Cota* and are included in the monophyletic, well-supported clade of *Cota* both in ITS and in the cpDNA BI trees (*clade b*, Figs. 2.1, 2.2). Taking into consideration their achenes morphology, Fedorov (1961) and Iranshahr (1986) had already pointed out their close relationship to *Cota* by classifying them within sect. *Cota*. Based both on molecular and morphological characters, we therefore propose for them the formal inclusion in the genus *Cota* (see Appendix A4). On the other side, two taxa formally included within the genus *Cota*, the annual, east-Mediterranean species *Cota halophila* and *Cota samuelssonii*, are quite distinct from the core of *Cota* both on molecular and morphological grounds. In the ITS-BI tree, the two species are included in well supported clades within *Anthemis* (*clade f* and *clade g1*, respectively, Fig. 2.1), and the same position is confirmed for *C. halophila* by the cpDNA dataset (*clade e&f*, Fig. 2.2b; for *C. samuelssonii* we lack plastid data). In the PCA of the micro-morphological characters (Fig.

2.6), these species cluster together with the other *Anthemis* species and are isolated from the representatives of *Cota*. They lack a distinctive character of the genus *Cota*, i.e. single, large crystals of calcium-oxalate and, as already Boissier (1875), Yavin (1972), Grierson and Yavin (1975), and Mouterde (1983) observed, *C. halophila* is characterized by achenes only slightly compressed and ± rombic or triangular in section. Yavin (1972), taking into account that it resembles species of section *Anthemis*, put *C. halophila* within the uni-specific series *Halophila* Yavin of the section *Anthemis*. Mouterde (1983), who included in his flora also *C. samuelssonii*, described the shape of its achenes as quadrangular, a character state that our micro-morphological analyses have confirmed. For these two species we, therefore, propose their transfer to the genus *Anthemis*.

Ammanthus – The four annual *Anthemis* species *A. tomentella*, *A. filicaulis*, *A. ammanthus* and *A. glaberrima*, all of them being endemic to Crete, were considered to represent an independent genus by Boissier (1849) due to the occasional absence of paleas. The observation of Greuter (1968) that this character expression is very variable and that, apart from this character, they resemble *Anthemis* in all other characters, led this author to include the species again into the broad generic concept of *Anthemis*. This view was backed by findings of Mitsuoka & Ehrendorfer (1972) who observed that the presence/absence of paleas is under oligogenic control in the whole tribe of Anthemideae. As a consequence, Yavin (1972) included all four species in her *A.* ser. *Ammanthus* (Boiss.) Yavin of the section *Anthemis*. On the other hand, however, Fernandes (1975, 1976) considered only two of the species (*A. tomentella* and *A. filicaulis*) distinct enough to merit subgeneric rank (*Anthemis* subg. *Ammanthus*), while the other two species were left unclassified in her treatment of the genus for *Flora Europaea*.

Our present analyses are quite equivocal about the phylogenetic relationships of the four representatives of the so-called '*Ammanthus*-group'. While in the ITS-BI tree only two of the four species (the subg. *Ammanthus* sensu Fernandes, 1975, 1976) are sister to each other with high statistical support (*clade g2*, Fig. 2.1) and the other species are also part of the *Anthemis* core clade, in the cpDNA-BI tree,

A. filicaulis, A. tomentella, and *A. ammanthus* (together with the Aegean species *A. scopulorum* and *A. rigida)* form a well-supported subclade in the *Cota* clade *(clade b,* Fig. 2.2a). Confirming rather the nrDNA-ITS than the cpDNA scenario, micro-morphological analyses reveal the closer relationships with the genus *Anthemis* and argue against an affinity to the genus *Cota* (Fig. 2.6). These incongruences are not restricted to the members of the *Ammanthus*-group, but also involve four other annual species being widespread in the Aegean region, *A. rigida, A. scopulorum, A. peregrina* and *A. tomentosa,* the latter three being members of the series *Tomentosae* Yavin (Geôrgiou, 1990). A possible explanation of these incongruences may be that both species groups were affected by transfer of chloroplast via hybridisation with a member of the genus *Cota.*

Infrageneric delimitation

The traditional infrageneric classification of *Anthemis* and *Cota* is not entirely reflected by our phylogenetic analyses. The sectional classification based on life-form seems to be quite artificial in both genera. Within the genus *Anthemis,* the perennial species, traditionally included in section *Hiorthia,* are located at least in three different clades (clades *a, d, g;* Fig. 2.3), while representatives of section *Anthemis,* which includes taxa that share the annual life form as unique synapomorphy, belong to many different evolutionary lineages. The same occurs within the genus *Cota,* where annuals (representatives of the section *Cota)* and perennials (formally included within section *Anthemaria)* are not forming separate clades (Figs. 2.1, 2.3).

Anthemis **sect.** *Hiorthia* – The divergences between the three evolutionary lineages within the traditionally circumscribed section *Hiorthia* are well supported by both ITS and cpDNA (Figs. 2.1, 2.2). While the isolation of the four basal perennial species *A. calcarea, A. marschalliana, A. trotzkiana* and *A. fruticulosa* was already discussed above, the morphological and phytogeographical differences between the two other perennial lineages *(clade d* on one hand and that included in *clade g* on the other) were comprehensively described by Oberprieler (2001). There is only one species, namely *A. cupaniana*, an endemic, rare, tetraploid species occurring in the mountains of Sicily, that shows a connecting position between the two lineages, as it is included with well support in *clade d* (cpDNA dataset; Fig. 2.2) and in *clade g2* (ITS dataset; Fig. 2.1). Fernandes (1975) mentioned the several affinities of *A. cupaniana* with *A. punctata* (*clade g2*, Fig. 2.1 and *clade g*, Fig. 2.2), a tetraploid species occurring in the mountains of North Africa (e.g. large and showy flower heads) but remarked the differences in the achenes, which are shorter and not tuberculated. On the other hand, the smoothed achenes resemble those of *A. cretica* subsp. *cretica* (*clade d*, Figs. 2.1, 2.2b), a tetraploid species with a wider range covering the mountains of South Europe and occurring also in Sicily. Thus, morphological characters support the genetical findings and argue for a hybrid origin of *A. cupaniana* from *A. punctata* and *A. cretica* subsp. *cretica*.

Anthemis* sect. *Chia – Due to several unique macro-morphological characteristics the annual, central-eastern Mediterranean species *Anthemis chia* was already included by Eig (1938) into a separate group (the group of *A. chia* Eig) and later formally incorporated within the monospecific section *Chiae* by Yavin (1972). Oberprieler (1998, 2001) found a peculiar micro-morphological characteristic that supported the uniqueness of *A. chia* and its divergence from other representatives of *Anthemis*, i.e. large myxogenic cells that disintegrate when soaked in water. Our present analyses validate only partially the segregation of *A. chia* from all other representatives of *Anthemis*: while in the ITS-BI tree, this species has a rather basal, but not supported position between *Anthemis* and *Cota* (Fig. 2.1), in the cpDNA-BI tree, *A. chia* is strongly included within the main core of *Anthemis* (Fig. 2.2b, *clade g*). The micro-morphological analyses, however, show that the distinctive character of large, disintegrating myxogenic cells is also shared by two atypical representatives of the genus *Anthemis* (*A. ammanthus* and *A. filicaulis*; see above), by *Nananthea* and by some members of *Tripleurospermum* (e.g. *Tripleurospermum parviflorum*). In the BEAST analysis, which includes all three markers, *A. chia* has a quite basal position (Fig. 2.3), but we can not exclude that this may be due to the 'soft' conflict among ITS and cpDNA datasets. Therefore, we can not support the elevation of this species to sectional or even subgenerical rang (as proposed by Oberprieler, 2001).

Anthemis* sect. *Odontostephana – The two annual species *A. odontostephana* and *A. tubicina* are distributed in the Middle East (probably reaching even northern India; Eig, 1938) and share many unique morphological characters, which have led different authors to treat them at least at sectional rank (Eig, 1938; Yavin, 1970). Our analyses weakly support the independence of this group of species from the main core of *Anthemis*: while in the PCA of the micro-morphological characters (Fig. 2.6) they cluster on the left side of the plot together with the outgroup taxa *Tripleurospermum* and *Nananthea*, both ITS and cpDNA data show a closer relationship with a group of at least 9 annual species (*clade e*, Fig. 2.1 and *clade e&f*, Fig. 2.2b). In the BEAST analysis, where all markers converged together, this group of species forms a well supported clade (*clade e*, Fig. 2.3) and *A. tubicina* (we lack plastid information for *A. odontostephana*) is the older and more basal species. Four of the species included in this clade (*A. edumea*, *A. deserti-syriaci*, *A. breviradiata* and *A. wettsteiniana*) were treated by Eig (1938) as member of the 'group of *A. wettsteiniana*' Eig and, later, by Yavin (1972) as members of the series *Melampodinae* Yavin of section *Anthemis*, as they all share carinate paleas, leaves with a distinct petiole and tuberculate-auriculate achenes. The group defined by Eig (1938) included also *A. homalolepis*, which however has in common with the other species only the shape of the achenes. Yavin (1972), therefore, excluded it from her ser. *Melampodinae* and treated it as member of section *Rascheyanae*. The inclusion in this section is however not clear, as our morphological analyses show that this species lacks the peculiar characteristic of the section, i.e. very slender achenes with an exceptional width to length ratio. Of the last three species included in *clade e* (Fig. 2.3), *A. susiana* and *A. mirheydari* (previously never included in any group or series) share with the other representatives of the group their distinctive characters (i.e. carinate paleas, pedunculate leaves and tuberculate-auriculate achenes), while *A. scariosa*, treated by Eig (1938) and Yavin (1972) as unique member of the homologous group ('the group of *A. scariosa*' Eig) or series (ser. *Scariosae* Yavin of sect. *Anthemis*), is peculiar in many characters (lanceolate paleas, hairy tubes of ray and disc florets, sessile leaves). The affinities of *A. tubicina* with both *A. scariosa* (in the lanceolate paleas) and the other species belonging to *clade e* (in the pedunculate leaves and the tuberculate-auriculate achenes) are therefore remarkable. We could suppose that *A. odontostephana* and *A. tubicina* have maintained some plesiomorphic characters that the derived species have inherited only in part. The quite isolated position in the PCA (Fig. 2.6) near the outgroup taxa and the wider geographical area occupied by these two species (*A. odontostephana* and *A. tubicina* are widespread from Syria to Afghanistan, Pakistan and even northern India, while all other species are characterized by a narrower range, from Palestine to western Iran) may corroborate this hypothesis. The narrowly distributed species may have originated from the widespread *A. odontostephana* and *A. tubicina* through a process of fragmentation (as for example in *Saxifraga*; Conti & al., 1999) probably related to the increasing aridity in these regions. As in the above-discussed case of *A. chia*, we can not support the elevation of *A. odontostephana* / *A. tubicina* to sectional rank, but we may remark the peculiarity of the whole clade that seems to assume an intermediate position between the core of *Anthemis* and the genus *Cota*.

Anthemis sect. Rascheyanae – Section *Rascheyanae* traditionally comprises c. 15 annual species widespread in the East (from Palestine to Syria, Iraq, Iran, Turkmenistan, Afghanistan, Pakistan) and that are characterized by very slender achenes. Preliminary molecular analyses based on nrDNA (Oberprieler, 2001) have shown a close relationship between the only two species included in the mentioned study. If we exclude *A. homalolepis* from sect. *Rascheyanae* as discussed above, our present analyses show that the 10 representatives included here are all members of a monophyletic clade, well supported both in the ITS-BI tree and in the combined BEAST analysis (*clade c*, Figs. 2.1, 2.3). The peculiar characteristic of this group seems to be rather macro- (slender achenes) than micro-morphological, as in the PCA of the micro-morphological characters the species belonging to this clade are not separated from other representatives of *Anthemis* (Fig. 2.6). We should here notice that members of section

Rascheyanae are grouped together with other species that are not formally included in the section. They are all annual species also distributed in the same geographical region and adapted to the same climatic conditions (Lo Presti & Oberprieler, 2009; see chapter 3) but, with the exception of some taxa (i.e. *Anthemis kandaharica*), they are not characterized by slender achenes.

Anthemis sect. Maruta – The other section of the genus *Anthemis* characterized by a clear morphological character, i.e. subulate paleas, is the section *Maruta*, which comprises 14 annual species (Yavin, 1970) mainly distributed in the East Mediterranean region. Our analyses, which include 9 taxa belonging to this section, can not confirm the monophyly of this group of species, as only in the ITS BI-tree six species form a well supported clade (within *clade f*, Fig. 2.1). In the cpDNA-BI tree (Fig. 2.2) they are all (with an exception, see below) included within *clade g* but they do not form a monophyletic group. The same result is observed in the combined phylogenetic analysis (Fig. 2.3) and in the PCA of the micro-morphological characters (Fig. 2.6). It is however interesting to note that the two perennial species *A. tigrensis* and *A. yemensis* share the distinctive character of the section, the subulate paleas, and are placed with the other members of the section in the well supported clade of the ITS-BI tree. They have the southernmost distribution of the whole genus *Anthemis* as they grow in the mountains of Yemen and Saudi Arabia (*A. yemensis*; Ghafoor & Al-Turki, 1997) or in the afro-alpine habitats of Ethiopia, Uganda and Kenya, reaching even 4,000 m (*A. tigrensis*; Hedberg, 1957; Agnew & Agnew, 1994; Tadesse, 2004). The derived position within the clade of both species suggests that the evolution of the perennial life form could be derived and may probably be correlated with higher maximum elevations accompanied by a shift to montane habitats, as already observed in other taxa (i.e. *Lupinus*; Drummond, 2008). Finally, it is noteworthy the position of the annual, western asiatic *A. microcephala*. While in the ITS-BI tree it groups together with the other representatives of the section, in the cpDNA-BI tree it is a well supported sister to the outgroup taxon *Tripleurospermum*. This fact suggests a hybrid origin of this species, also supported by the micro-morphological analyses, where *A. microcephala* assumes an intermediate position between *Tripleurospermum* and the main group of *Anthemis*. Successful crossing experiments between *Tripleurospermum* and another member of sect. *Maruta, A. cotula* (Kay, 1965; Mitsuoka & Ehrendorfer, 1972), further corroborate this hypothesis.

Time of divergence and life history variation

The estimated ages of the genera *Anthemis* and *Cota* are based on four independent calibration points and are derived from one nuclear (ITS) and two chloroplast (*psbA-trnH* and *trnC-petN*) markers. They generally agree with previously inferred ages for representatives of the tribe Anthemideae solely based on ITS data with up to two external calibration points and obtained via non-parametric rate smoothing (Oberprieler, 2005) or penalized likelihood (Lo Presti & Oberprieler, 2009; see chapter 3). In particular, the age of the tribe Anthemideae (24.5 ± 7.8 Ma) is roughly in line with other molecular dating studies in Compositae (Wikström & al., 2001; Wagstaff & al., 2006) and matches the range in which all of the known major radiations of the family occurred (35-25 Ma; Funk & al., 2009).

The LTT plot for the major clades shows different rates of lineage branching. While *clade g* and *clade b* show a constant increase of lineage in the last 10 Ma, the radiations of *clade d* (*A. cretica* clade) and *clade c* (*A. rascheyana* clade) have occurred mainly in the last 2 Ma. These two clades are adapted to quite different habitats: while members of *clade d* are perennial species widespread in the mountains of the whole circum-Mediterranean region, representatives of *clade c* are annual species adapted to the arid environments of the Arabian Peninsula and western Asia. However, their more or less contemporary radiation may be related to the same climatic event: the glaciations of the Quaternary. On one side, glacial-interglacial cycles have promoted range expansion followed by isolation of high-altitude plants, a process well documented for many alpine groups (see, for example, Comes & Kadereit, 2003; Vargas, 2003; Koch & al., 2006, Tribsch & Schoenswetter, 2003) and also plausible for *clade d*. On the other side, the global cooling of the Pliocene caused an increasing of aridity in eastern Africa (deMenocal, 1995), the Arabian Peninsula and western Asia

(Clemens & al., 1996). As a consequence, the vegetation shifted from closed canopy to open savannah vegetation (deMenocal, 1995; Thompson, 2005), determining the evolutionary success of an arid-adapted flora (see Crisp & al., 2009), as it may be supposed for *clade c*.

The two genera *Anthemis* and *Cota* experienced their radiations in the last 10 Ma (Late Miocene): in the Mediterranean region this period is characterized by considerable climatic changes, as a trend toward progressive aridification started (Ivanov & al., 2002; Fortelius & al., 2006; Van Dam, 2006). The Late Miocene coincides also with the first appearance of the annual life form (Wolfe, 1997), which is considered to be promoted as one possible adaptation of plants to arid environments (Datson & al., 2008; Evans & al., 2005; Hellwig, 2004; Verboom & al., 2003; Fiz & al., 2002; Schaffer & Gadgil, 1975). Within *Anthemis* and *Cota*, the annual life form has probably evolved several times, as it occurs in almost all major clades (Fig. 2.3). Polyphyly of annual species was also found in other genera or subtribes, as for example in Centaurineae (Hellwig, 2004), *Veronica* (Albach & al., 2004), *Saxifraga* (Conti & al., 1999) or *Astragalus* (Liston & Wheeler, 1994). But, intriguingly, also the perennial life form occurs in many separate clades, thus arguing for independent evolutionary lineages. Therefore, the question arises whether the perennial life form is always to be considered plesiomorphic or it could also develop as secondary adaptation from annual histories (as, for example, in *Lupinus*, Drummond, 2008; *Trifolium*, Ellison & al., 2006; or *Medicago*, Bena & al., 1998). Annuals and perennials are characterised by different reproductive strategies: whereas perennials reproduce more than once per lifetime, annuals allocate all resources in the reproduction and then die (Drummond, 2008; Evans & al., 2005). Therefore, it is generally acknowledged that the evolution from annual to perennial would be complicated by the fact that the plant should 'learn' how to allocate resource for survival functions (see Bena & al., 1998). However, the shift to the perennial form may be the result of the selective advantages linked to the different ecological factors that characterized montane habitats: a more stabile environment without seasonal fluctuations might result in an enhancement of adult survival (Drummond, 2008) and hence trigger the perennial against the annual life form. We can not therefore exclude that, near basal, plesiomorphic perennial species (as, for example, the four basal species *A. calcarea*, *A. marschalliana*, *A. trotzkiana* and *A. fruticulosa*), other perennials may have secondary originated from annuals (as it could be the case for the species distributed in the mountains of North Africa: *A. abylaea*, *A. pedunculata*, *A. punctata*).

The rapid increase in the number of taxa of both genera in the last 10 Ma, related to the factors discussed above, may furthermore result in the lack of well-supported basal nodes both in ITS and in cpDNA datasets, as it was already proposed for other plant groups such as *Bomarea* (Alzate & al., 2008), *Inga* (Richardson & al., 2001), *Lupinus* (Hughes and Eastwood, 2006) or *Costus* (Kay & al., 2005). Therefore, as proposed by Alzate & al. (2008), alternative sources of phylogenetic information such as single copy nuclear genes or even mitochondrial DNA, may shed light to the first stages of the radiation of *Anthemis* and *Cota*.

Acknowledgments

This research was supported by the Bavarian Research Foundation (BFS), by the German Academic Exchange Service (DAAD), by the German Research Foundation (grant OB 155/7-1) and by the SYNTHESYS project of the EU to RMLP and CO (AT-TAF-1618, AT-TAF-1731). We would like to thank P. Hummel for his technical support in the laboratory at the University of Regensburg and all the keepers of the Herbaria visited (B, BC, M, S, W, WU).

Chapter 3

Evolutionary history, biogeography and eco-climatological differentiation of the genus *Anthemis* L. (Compositae, Anthemideae) in the circum-Mediterranean area

Materials and methods

Plant material, sequencing and phylogenetic reconstructions

Our analyses involved 141 of the 188 species belonging to the formerly congeneric, closely related genera *Anthemis* and *Cota* (see Appendix A1). For the 168 accessions sequenced, 30 ITS sequences came from former published accessions (Oberprieler, 2001; Oberprieler & Vogt, 2006), the others were newly obtained from herbarium specimens held in B, W, WU and S. Based on the results of Oberprieler & al. (2007), 19 representatives of the tribe Anthemideae were used as outgroup taxa.

The phylogenetic reconstruction presented here is based on a single nuclear marker, the nrDNA ITS region. There are problems related to the use of a single nuclear marker in general and of the ITS region in particular: in general, a single marker results in less resolution than a multiple marker, but, more critically, may miss reticulate evolution due to hybridization and polyploidization (Holland & al., 2004; Paun & al., 2005), since it can effectively give a gene tree. Secondly, as the nr DNA ITS region contains very numerous tandem repeats, there is the possibility of finding different paralogous copies of the region in the same individual (Álvarez & Wendel, 2003). If these are mixed, they might lead to incorrect phylogenetic reconstructions (Bailey & al., 2003). Despite these concerns, however, we have concentrated on an extensive taxon sampling than

on an extensive character sampling, as (a) for biogeographical and eco-
climatological reconstructions a (nearly) complete taxon sample is preferable,
and (b) extensive character sampling at the cost of limited taxon sampling can
not safeguard phylogenetic reconstructions against artificial results (Soltis &
Soltis, 2004). Polyploidisation is observed only in the *A. cretica*-group (i.e.,
A. sect. *Hiorthia*), a closely-knit assemblage of around 24 perennial species.
Moreover, results of phylogenetic reconstructions based on the two chloroplast
regions *psbA-trnH* and *trnC-petN* for the genera *Anthemis* and *Cota* (see chapter
2) do generally agree with patterns seen in the nrDNA ITS phylogeny.
Therefore, we feel confident that the ITS-based phylogeny reflects a taxon- and
not a gene-phylogeny.

DNA was extracted from 20-30 mg crushed leaf materials according to a
modified protocol based on the method of Doyle & Doyle (1987). Amplification
of nrDNA ITS1 and ITS2 was performed using the following primers: 18SF and
26SR (Rydin & al., 2004), ITS5A (Funk & al., 2004), ITS2, ITS3, ITS4 (White
& al., 1990). Amplification reactions followed Oberprieler & al. (2007). The
polymerase chain reaction (PCR) products were purified with QIAquick PCR
Purification Kit (Qiagen, Hilden, Germany) or with Agencourt AMPure
magnetic beads (Agencourt Bioscience Corporation, Beverly, Massachusetts,
USA). Cycle sequencing reactions were performed using the DTCS Sequencing
kit (Beckman Coulter, Fullerton, California, USA), following the manufacturer's
manual. The fragments were separated on a CEQ8000 sequencer (Beckman
Coulter, Fullerton, California, USA). The DNA sequences obtained were
carefully checked for presence/absence of polymorphic sites. We used the
IUPAC (International Union of Pure and Applied Chemistry) ambiguity codes to
indicate single nucleotide polymorphisms. A site was considered polymorphic
when more than one peak was present on the electropherogram, with a minimal
intensity of 25 % for the weakest peak compared to the strongest signal (Fuertes
Aguilar & al., 1999; Mansion & al., 2005). The observation that these
polymorphisms were very rare in the directly sequenced PCR products (0.46%
of the total number of sequenced position) allowed us to abstain from cloning of
products. Sequences were aligned using Clustal W (Thompson & al., 1994) as

implemented in BioEdit version 7.05.2 (Hall, 1999) and the alignment was optimized manually. Gaps were treated as missing data.

The maximum parsimony (MP) analysis of the data set was performed using the heuristic search algorithm of PAUP* 4.0 version beta 10 (Swofford, 2002), with ACCTRAN, MULPARS, tree bisection–reconnection (TBR) branch swapping, and 1000 random addition sequence replicates. Character states were treated as unordered and unweighted. To test the robustness of individual branches, bootstrap (BS) values (Felsenstein, 1985) were calculated. The bootstrap analysis was performed using 100 replicates, simple addition sequence, ACCTRAN, TBR branch swapping, MULPARS and saving 10 trees per replicate.

To carry out a maximum likelihood (ML) analysis, the best-fitting model of sequence evolution was determined using a hierarchical likelihood ratio test implemented in Modeltest version 3.06 (Posada & Crandall, 1998). This resulted in the acceptance of the model of Tamura & Nei (1993) with a gamma distribution of substitution rates over the sites (TrN+Γ). Using this model, the ML search was performed with Treefinder version June 2004 (Jobb & al., 2004). ML bootstrap analysis with 1000 replicates was also carried out using the same software.

Finally, a Bayesian inference (BI) phylogenetic analysis was performed with MrBayes version 3.1.2 (Ronquist & Huelsenbeck, 2003), implementing a GTR model with gamma distribution ('Lset NSt=6 rates=gamma'), without fixing rates and nucleotide frequencies, as these parameters are estimated from the data during the analysis. The analysis was conducted using four chains and was run for a total of 5×10^6 generations, with trees sampled every 100th generation. To ensure approximation to stationarity, we checked that the average standard deviation of split frequencies was approaching zero, as suggested by Ronquist & al. (2005). From the resulting 5×10^4 trees, the first 3,000 trees sampled before the likelihood values converged in a maximum value were discarded. A majority rule consensus tree of the remaining 47,000 trees was computed.

Chronogram construction and calibration

As no fossils are known for *Anthemis* or closely related taxa, we have calibrated the chronogram with two constraints. First, we used the maximum age of the Canary island Tenerife (11.6 Ma; Juan & al., 2000) to constrain the maximum age of the split between the Mediterranean representatives of the clade and the monophyletic group of *Gonospermum canariense* + *Lugoa revoluta,* endemic to the Canary Islands Tenerife, Gomera, La Palma and El Hierro (Greuter *et. al,* 1984-1989). Secondly, we constrained the split between *Anthemis/Cota* and *Anacyclus* to a minimum age of 14.1 Ma based on the externally dated tree of the tribe Anthemideae (Oberprieler, 2005), which was obtained using fossil evidence for the whole family of the Asteraceae (Graham, 1996).

With these constraints, we performed a penalized likelihood analysis (PL; Sanderson, 2002) with a truncated Newton algorithm, applying r8s version 1.70 (Sanderson, 2003) to the fully resolved ML phylogram. We have used the cross-validation approach in r8s to infer the optimal level of rate smoothing for our data. The smoothing parameter found to minimise the cross-validation score (in the form of a raw sum of squared deviations) was 0.05, chosen on a scale from 0.001 to 100. This value was then used (1) in the estimation of divergence times in the phylogram, and (2) in the estimation of the level of error in divergence time estimates. To prevent the optimization algorithm from converging in a local optimum, the PL searches were started at ten different initial time estimates and were restarted ten times for each round (Paun & al., 2005). As suggested by Sanderson (2004), we allowed the collapse of those internal branch lengths approaching to zero ('collapse' command) to prevent problems in the calculations. To receive estimates of the error rate associated with divergence times, we first generated 1000 phylograms in PAUP* (Swofford, 2002) from bootstrap resampling of the original dataset. We then summarized nodal ages obtained from all trees (using the 'profile' command in r8s) and used the resulting standard deviations to calculate the 95% confidence intervals for the estimated ages of the original ML phylogram.

In order to assess the temporal pattern of diversification within *Anthemis* s.l., the logarithm of the cumulative number of lineages was plotted against the absolute date of each node obtained using PL dating (lineages-through-time plot, LTT). This type of graph gives a coarse impression of the temporal diversification of lineages and is considered to allow an assessment of climatic and/or geological influences on speciation rates when (and only when) a constant extinction rate is assumed (Barraclough & Reeves, 2005; Kadereit & Comes, 2005; Savolainen & Forest, 2005).

Distribution data

General information for distribution patterns of the terminal taxa came from a personal inventory based on Euro+Med Plant Database (Greuter & al., 1984-1989) and floras of areas not covered in the Euro+Med project: Iran (Iranshahr, 1986), USSR (Fedorov, 1961), and Saudi Arabia (Ghafoor & Al-Turki, 1997). Synonyms were checked and a new, complete check-list of species belonging to the genera *Anthemis* and *Cota*, comprising 195 species, was compiled.

To carry out the eco-climatological reconstruction, we collected 4,151 distribution points (1 to 369 locations per species with an average of 26 localities per species) for 147 of the 150 species included in the analyses. For three species (*C. dubia*, *C. macroglossa* and *C. lincezveskyi*), no precise information was found and these were therefore excluded from all subsequent analyses. Distribution data came from the following data sources:

(1) In total, 3,583 records were obtained by collecting the information found in the previously checked and/or identified herbarium specimens held in B, BC, M, W, WU (Holmgren & Holmgren, 1998).

(2) Other sources of information were the monographs or floras with detailed occurrence data for each species (Boissier, 1875; Eig, 1938; Tadros, 1953; Fedorov, 1961; Yavin, 1972; Grierson & Yavin, 1975; Feinbrun-Dothan, 1978; Mouterde, 1983; Iranshahr, 1986; Ghafoor & Al-Turki, 1997; Oberprieler,

1998), which led to an additional 551 records. All of these 4,134 records were then georeferenced with the aid of GoogleEarth beta version 4.0.002 (openGL available at http://earth.google.com/) and GNS (Geographic Names Search - http://gnswww.nga.mil/geonames/GNS/index.jsp).

(3) Finally, the Global Biodiversity Information Facility (accessed through GBIF Data Portal, www.gbif.org, 2007-03-06) provided information for 10 species [17 Occurrence Data (OD) records provided by: Lund Botanical Museum, Sweden (10 OD records, 5 species); Vascular Plant Herbarium (MA), Spain (3 OD records, 3 species); Herbarium RNG, School of Plant Sciences, The University of Reading, UK (2 OD records, 1 species); Botanischer Garten Jena, Germany (1 OD record); Jardín Botánico de Córdoba – Spain (1 OD record)].

Prior to subsequent analyses, all points were plotted in geographical space to ensure no outliers were processed (Graham & al., 2004). Outlying points at species level were identified based on the previously compiled check-list and either verified or removed.

Biogeographical analyses

The 11 areas used in the biogeographical analyses were defined on the basis of two criteria: (1) geographical boundaries that may have acted as barriers to dispersal (e.g. water bodies, mountain ranges); and (2) congruent distributional range (sympatric distribution) shared by two or more species ('areas of endemism', Platnick, 1991; Sanmartín, 2003; Oberprieler, 2005), chosen by overlapping the distributional ranges of species as derived from polygons ('convex hulls') constructed for each species around all of its point localities, omitting species with only one or two records. An 'area' was defined when shared by five, non-widespread species, a number that represents a compromise between too large and too small areas and thus allowing the recognition of disconnected areas. The areas obtained are shown in Fig. 3.1.

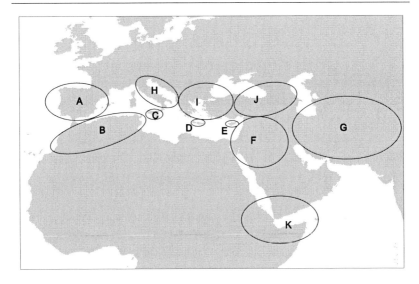

Fig. 3.1. Areas adopted in the DIVA analysis. A: Iberian Peninsula. B: Morocco, Algeria, Tunisia. C: Sicily. D: Crete. E: Cypro. F: North Arabian Peninsula and Sinai. G: Iran and eastern Asia. H: Italy and Balkans. I: Aegean region and western Turkey. J: eastern Turkey and Caucasian region. K: South Arabian Peninsula and eastern Africa. (Base map modified from a freely distributed global map of the world downloaded at www.diva-gis.org/Data.htm).

Many different methods are available to reconstruct ancestral states in a parsimony or in a model-based scenario. The dispersal–vicariance analysis (DIVA version 1.1; Ronquist, 1996, 1997), an event-based parsimony method, assigns no costs to vicariance events, thus explicitly favouring vicariance against dispersal. This property of DIVA reconstruction is intentionally favoured in the present study because vicariance events are more easily pinpointed to a specific point in time (the speciation event leading to two daughter lineages) than dispersal events, which occur along branches of a phylogenetic tree. This allows a sliding-window analysis to measure the influence of geographical forces on the diversification of the study group. Secondly, while vicariance events are directly connected to geological processes (e.g., loss of land connections), dispersal events may be related to them (short-distance dispersal over land bridges) but need not be (long-distance dispersal). Thirdly, we consider the reconstruction of vicariance events to measure geographical mobility through time analogous to the method adopted for the eco-climatological differentiation through time ('climatic vicariance', see below). DIVA also allows for reconstruction of ancestral areas without *a priori* assumptions concerning relationships between areas (Ronquist, 1997; Mansion & al., 2008). This is particularly appropriate in regions such as the Mediterranean region characterized by a highly reticulate geological history due to migrating island arcs, fragmentation of tectonic belts, and the recurrent cycles of marine regressions and transgressions during the Cenozoic (Sanmartín, 2003; Oberprieler, 2005). For all these reasons, we adopted DIVA as the method to reconstruct the ancestral distributions.

As the total number of taxa was larger than the maximum number allowed by the DIVA software, we conducted two separate, nested DIVA analyses on the fully resolved ML tree. The result obtained for the root node of the first analysis on a pruned twig of the total tree was treated as a supplementary taxon in the more inclusive analysis, which comprised the more basal part of the tree. DIVA reconstructions of ancestral distributions were conducted twice: first without restricting the number of areas for each node, and second, restricting it to a maximum of two areas. The latter was based on the arithmetic mean of 1.9 occupied areas over the numbers of areas occupied by each extant species (see Oberprieler, 2005).

As DIVA currently does not calculate the relative support for alternative character states and does not account for uncertainty in phylogenetic inference, we also chose to compare this method with a 'presence coding' likelihood optimisation (as implemented in Mesquite version 2.0; Galley & al., 2007; Maddison & Maddison, 2007). 'Presence coding' (Hardy & Linder, 2005; Hardy, 2006; Galley & al., 2007) is a way to code each state as a separate presence–absence character. In contrast to DIVA, this likelihood-based method uses branch length information to calculate the probability of character state change (Ree & al., 2005), which, on a chronogram, is directly related to time (Galley & al., 2007). Even if it accounts for polymorphism both in terminal and in ancestral nodes, it has the disadvantage of artefactual reconstruction of 'no-state' at some internal nodes as a consequence of the assumption of character independence

Table 3.1. Bioclimatic variables (i.e. biologically meaningful variables derived from the monthly temperature and rainfall values) with results of the tests for the phylogenetic signal and the factor loadings from the principal components analysis (PCA) of 322 operational taxonomic units (OTUs) consisting of 162 sampled species (terminal taxa) and 160 reconstructed ancestors (internal nodes). In the *K*- and QVI-tests the strength of the phylogenetic signal is measured by the *K*- and QVI-statistics, respectively, with *P* describing the significance of each statistic.

Bioclimatic variables	Meaning	*K*-test		QVI-test		PCA analysis	
		K statistic	*P*	QVI statistic	*P*	PC1 loadings	PC2 loadings
BIO_1	Annual mean temperature	0.012	0.000	0.49	0.001	0.928	0.256
BIO_2	Mean diurnal range (Mean of monthly (max temp - min temp))	0.005	0.000	0.564	0.001	0.445	-0.655
BIO_3	Isothermality (BIO_2 / BIO_7 * 100)	0.008	0.041	0.579	0.001	0.471	0.228
BIO_4	Temperature seasonality (standard deviation *100)	0.007	0.019	0.531	0.001	-0.081	-0.832
BIO_5	Max temperature of warmest month	0.007	0.001	0.534	0.001	0.851	-0.319
BIO_6	Min temperature of coldest month	0.012	0.000	0.528	0.001	0.68	0.677
BIO_7	Temperature annual range (BIO_5 – BIO_6)	0.005	0.016	0.544	0.001	0.091	-0.886
BIO_8	Mean temperature of wettest quarter	0.004	0.023	0.587	0.001	0.46	0.243
BIO_9	Mean temperature of driest quarter	0.027	0.000	0.548	0.001	0.877	0.126
BIO_10	Mean temperature of warmest quarter	0.007	0.004	0.524	0.001	0.907	-0.079
BIO_11	Mean temperature of coldest quarter	0.015	0.000	0.493	0.001	0.795	0.549
BIO_12	Annual precipitation	0.006	0.004	0.571	0.001	-0.764	0.51
BIO_13	Precipitation of wettest month	0.004	0.019	0.657	0.001	-0.441	0.701
BIO_14	Precipitation of driest month	0.026	0.000	0.469	0.001	-0.869	-0.119
BIO_15	Precipitation seasonality	0.013	0.000	0.493	0.001	0.824	0.235
BIO_16	Precipitation of wettest quarter	0.005	0.006	0.649	0.001	-0.476	0.689
BIO_17	Precipitation of driest quarter	0.023	0.000	0.474	0.001	-0.882	-0.065
BIO_18	Precipitation of warmest quarter	0.018	0.001	0.497	0.001	-0.839	-0.052
BIO_19	Precipitation of coldest quarter	0.005	0.002	0.656	0.001	-0.282	0.666

(Mickevich & Johnson, 1976; Hardy & Linder, 2005; Hardy, 2006). For this reason, this method was chosen as a test for support: we conducted the analysis (1) on the ultrametric ML tree, and (2) on a sample of 1,000 trees obtained from the Bayesian analysis and transformed into chronograms. The support for each area reconstruction for each node was calculated as the product of topological uncertainty (measured as the posterior probability) and of area optimisation uncertainty (measured as the average of likelihood values over 1,000 trees).

Climatic data and phylo-ecological analyses

The data for current climatic conditions were obtained from the WorldClim dataset (www.worldclim.org), which contains records for the 1960-90 period with a 2.5 min (approximately 5 km^2) resolution (Hijmans & al., 2005). This dataset includes 19 'bioclimatic variables', i.e. biologically meaningful variables derived from the monthly temperature and rainfall values (see Table 3.1). The bioclimatic variables represent annual trends (e.g., mean annual temperature, annual precipitation), seasonality (e.g., annual range in temperature and precipitation), and extreme or limiting environmental factors (e.g., temperature of the coldest and warmest month, and precipitation of the wet and dry quarters). All these variables were used to describe an eco-climatological niche for each taxon under study, recording the mean, minimum, maximum and standard deviation for each bioclimatic variable for each terminal taxon.

For the reconstruction of eco-climatological niches of internal nodes, mean, minimum and maximum values of each climatic variable were independently optimized as continuous characters across the ML tree ('MaxMin coding', Hardy & Linder, 2005; Yesson & Culham, 2006) with the squared change parsimony optimization implemented in Mesquite version 2.0 (Maddison & Maddison, 2007). Since species are ecologically variable, polymorphy in the ancestors should be reconstructed, and MaxMin coding is a suitable method for obtaining ranges of values, as it does not require the discrete coding of continuous variables with its consequence of loss of information (Hardy & Linder, 2005).

For each optimized variable, the phylogenetic signal was tested by adopting two kinds of tests. The '*K*-test', implemented for continuous characters in the program PHYSIG.m version 2.0 (Blomberg & al., 2003), is based on a randomization procedure and compares the fit of the data to the tree with the fit obtained when the data are randomly permuted across the tips of the tree

(Blomberg & al., 2003). It provides information on the test statistic K which describes the strength of the phylogenetic signal and on its significance P. As the original branch lengths obtained from the ML analysis did not allow the computation of the test, rescaled branch lengths (original lengths x 100) were used, as suggested by the authors of the test (T. Garland, pers. comm.), and 1,000 permutations were run. We also computed the QVI (Quantitative Convergence Index; Ackerly & Donoghue, 1998), a different way of testing for phylogenetic signal in continuous characters based on linear parsimony algorithms. The QVI is equivalent to 1 minus the retention index (Farris, 1989). To perform this test, the software Cactus version 1.13 (Schwilk, 1999-2001; Schwilk & Ackerly, 2001) was adopted and 1,000 randomizations were performed.

A principal components analysis (PCA) was carried out based on mean values of all the eco-climatological variables for each terminal taxon and each internal node to obtain an interpretable two-dimensional ordination of taxa according to their niche similarities. In order to describe the overlap of eco-climatological niches of sister groups, we took the following steps. Values of each variable were normalised and a 19-dimensional hypervolume (H) was computed for each taxon x as the product of ranges for all 19 variables:

$$H_x = \prod_{i=1}^{19}(\max_{xi} - \min_{xi})$$

The pairwise overlap (HO_{ab}) among hypervolumes of sister-groups a and b for the 19 variables was calculated with the following formula:

$$HO_{ab} = \prod_{i=1}^{19}[\min(\max_{ai}, \max_{bi}) - \max(\min_{ai}, \min_{bi})]$$

When one factor resulted in a negative value (no overlap), HO was set to 0. Positive HO values were finally converted into a percentage value (HP) by dividing the size of the hypervolume intersect by the volume of the smaller of the two hypervolumes involved:

$$HP_{ab} = HO_{ab}/\min(H_a, H_b) \times 100$$

As a consequence, the total overlap ranged between 0% (no overlap at all, an event that we call 'climatic vicariance') and 100% (the smaller hypervolume being completely included in the larger one).

Integration of temporal, biogeographical and phylo-ecological analyses

All the results obtained from each separate analysis (molecular dating, reconstruction of ancestral areas, reconstruction of ancestral eco-climatological niches) were combined in the following calculations, with the aim of measuring the influence of geographical vs. eco-climatological forces on the diversification of the study group. Firstly, in order to analyse whether there are differences among clades concerning the relative importance of geographical vs. eco-climatological speciation events, we computed clade-wise proportions of both event classes (number of vicariant nodes per clade/total number of nodes in the clade x 100) and compared these values to the grand mean calculated for the whole tree.

Additionally, the temporal change in the relative proportions of geographical vs. 'climatic vicariance' events was estimated by using a sliding-window technique. Taking into account that the average of the 95% confidence interval (CI) of the divergence time estimates was 4.70 Ma, the size of the time window was set to 5 Ma and the slide increment to 0.1 Ma. In each of the resulting time-slices, we counted how many of the nodes included in each interval showed either a geographical or an eco-climatological vicariance event (number of vicariant nodes/total number of nodes x 100), and the resulting values were plotted against time.

Next two pages: Fig. 3.2. Dated phylogenetic tree from a maximum likelihood (ML) analysis of nrDNA ITS sequence data for *Anthemis* s.l. Polytomies in the chronogram are a consequence of the collapse in the calibration method of branches approaching zero-length. Values at the branches indicate bootstrap support values (below, regular), posterior probabilities (per cent) from the Bayesian analysis (below, *italics*) and bootstrap support values from the parsimony analysis (above, bold). (a): lower part of the tree. (b): upper part of the tree.

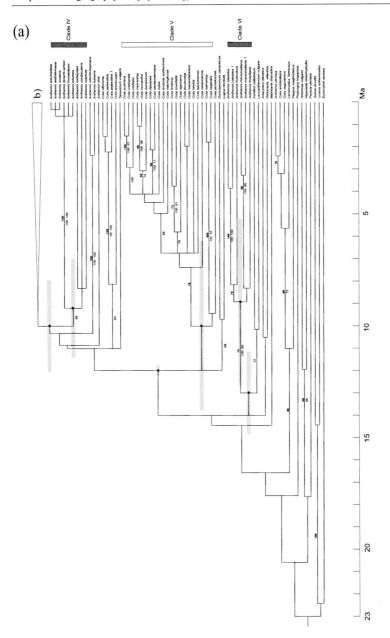

(b)

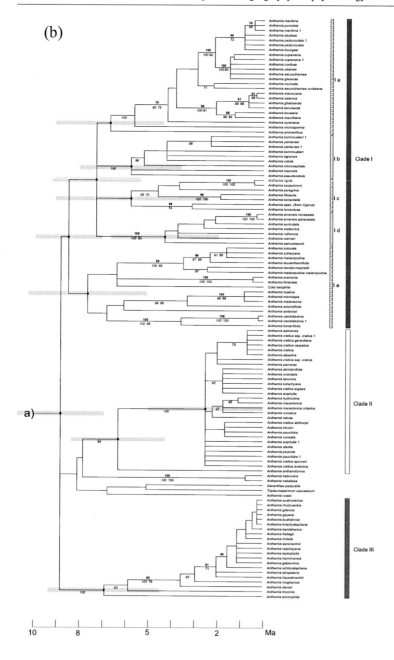

Results

Phylogeny and calibration

The alignment of all the 184 nrDNA ITS sequences is 525 bp long (282 for ITS1 and 243 for ITS2), with 329 variable positions including 230 parsimony informative substitutions. The ML search with no enforced molecular clock yielded a fully resolved tree (-lnL=6868.3835). The tree depicted in Fig. 3.2 shows the ML tree with branch lengths adjusted according to the results of the penalized likelihood calibration. As a consequence of the collapsing of branches that are resolved but approach the length of zero, polytomies in the chronogram are observed. The trees resulting from the BI and MP approaches are very similar to the tree depicted in Fig. 3.2 and are therefore not shown. The posterior probabilities of the BI analysis are indicated along the branches together with support values from bootstrapped ML and MP analyses. Divergence time estimates and 95% confidence intervals for relevant nodes, spanning around an average of 4.8 Ma, are shown in Table 3.2. The mutation rate obtained from our data (0.00468 mutations/site/Ma) falls inside the range available for taxa belonging the Asteraceae (between 0.00251 and 0.00783 mutations/site/Ma; Kay & al., 2006).

The analysis of nrDNA ITS data shows that the genus *Anthemis* s.l. is not monophyletic. The attempt to make it monophyletic by considering *Cota* a separate genus (Greuter & al., 2003) is not sufficient to yield monophyletic genera, as there are many other *Anthemis* species that fall outside the main clade of *Anthemis* s.s. (clade I – Fig. 3.2b). We have chosen to treat the largest clade that does not comprise members of the two genera *Tripleurospermum* and *Nananthea* as 'Anthemis s.s. group'. This group of species (clade I) is formed by 55 annual and perennial species and is widespread in the Mediterranean area. While it is not well-supported by the bootstrap analyses, it is possible to identify many well-supported sub-clades within it: clade Ia comprises 20 annual and perennial species occurring in the western part of the Mediterranean; clade Ib encompasses 7 annual species distributed in the eastern Mediterranean region and Arabian Peninsula; and clade Id contains 6 species widespread in the central Mediterranean.

Table 3.2. Summary of temporal, geographical and eco-climatological reconstructions. Stem and crown ages, geographical (DIVA area) and climatic (PCA plot panel) reconstructions of the ancestor of the clade, percentage of geographical and climatic vicariant nodes are indicated for each clade; "Mean" refers to the percentage of all vicariant nodes in the entire tree. The names of the clades are the same as in Fig. 3.2.

Clade	Stem age (Ma) (a)	Crown age (Ma) (a)	Geographical reconstruction (b)	% of geographic vicariant nodes in clade	Climatic reconstruction (c)	% of climatic vicariant nodes in clade
I	8.83 (± 1.87)	8.50 (± 1.44)	FJ/IJ	16.13	I	35.48
Ia	7.30 (± 1.90)	6.70 (± 1.84)	BD/DF	18.18	I	40.91
Ib	7.30 (± 1.90)	5.80 (± 2.20)	F/I/FI	25	I	25
Ic	7.30 (± 1.90)	5.80 (± 2.25)	FI	16.67	III	16.67
Id	7.30 (± 1.90)	4.30 (± 2.33)	I	16.67	I	16.67
Ie	8.50 (± 1.44)	7.67 (± 2.55)	J	6.25	I	56.25
II	6.33 (± 2.04)	2.50 (± 2.48)	I	37.93	III	17.24
III	8.83 (± 1.87)	6.90 (± 2.42)	IJ	20	I	60
IV	10.00 (± 2.05)	9.20 (± 2.21)	F	12.5	IV	62.5
V	12.00 (± 0.23)	10.00 (± 4.28)	HJ/IJ	11.11	III	22.22
VI	13.00 (± 2.34)	8.91 (± 3.72)	J	0	III	66.67
Mean				19.38		34.38

(a): Mean age estimates and 95% confidence intervals obtained with r8s (Sanderson, 2003).

(b): Reconstructions of ancestral areas according to DIVA analysis (code areas in Fig. 3.1).

(c): Panel numbers as in PCA-plot of Fig. 3.5.

Additionally, throughout the remaining tree it is possible to recognize further groups that are well supported in all the analyses. Clade II encompasses 20 perennial species distributed in the mountains of the circum-Mediterranean region; clade III comprises 21 annual species widespread in western Asia (Iran, Turkmenistan, Afghanistan, Pakistan); and clade IV includes 9 annual species occurring in the northern part of the Arabian Peninsula. A '*Cota*-group' (clade V), comprising almost all the species belonging to this genus, can be identified too.

However, only 11 of the total 22 species of this clade are well supported as a monophyletic group, and five other species belonging to the genus *Cota* are found in the previously mentioned clades together with *Anthemis* species. Finally, it is possible to recognize a very well supported basal group (clade VI) comprising 4 perennial *Anthemis* species endemic to the Caucasian region.

The LTT plot for *Anthemis* s.l. (Fig. 3.3) shows a constant increase of lineages, thus illustrating that the speciation rate remains higher than the extinction rate until present. A higher rate of lineage branching can be identified at around 10-8 Ma, a period that corresponds to the main radiation of the genus into the three clades I, II and III.

Biogeographical analyses

In the DIVA analysis with ancestral distributions constrained to a maximum of two areas, a total of 193 dispersal events were required. The results are the same as those shown by the unconstrained analysis (that required 218 dispersal events), except for the reconstruction at the very basal nodes, where a restricted ancestral distribution was not identified and all 11 areas were reconstructed as equally parsimonious. However, the difference between the two analyses is most likely due to the inclusion of clade Ib in the analyses, containing the widespread *A. cotula* L. and eight other accessions: when excluding this group from the unconstrained run, the same results for the basal nodes are produced by the two analyses.

Because the results obtained from the likelihood optimisation on the ultrametric ML tree are consistent with those observed in DIVA reconstructions (and are therefore not shown), we feel confident that the reconstruction of ancestral areas is independent from the method adopted. However, the analysis based on the 1,000 ultrametric MB trees (see Appendix A5) indicates that these biogeographical reconstructions are heavily impaired by the topological uncertainty at the basal nodes (see Fig. 3.2). The general biogeographical scenario of an east–west expansion presented below seems to be not perturbed too drastically by these flaws, as for older ancestors an exclusively eastern Mediterranean distribution is supported (e.g. ancestors in clades VI and V).

In the constrained search of DIVA, 32 nodes of the tree are reconstructed as vicariance events (black dots in Fig. 3.4). Even though there are many alternative solutions (J/FJ/HJ/IJ; see Fig. 3.4a) for the basal node of *Anthemis* s.l., all reconstructions suggest a Near East distribution of the ancestor. The common

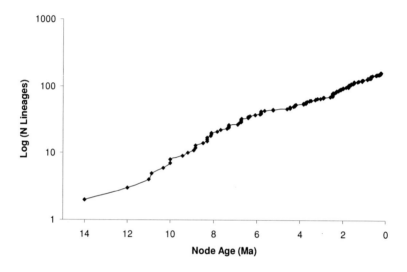

Fig. 3.3. Lineages-through-time (LTT) plot for the whole phylogeny of *Anthemis* s.l. from a maximum likelihood (ML) analysis of nrDNA ITS sequence data, using dates obtained by penalized likelihood dating method as implemented in r8s.

ancestor of clade VI seems to have originated in the eastern Anatolia–Caucasian region (J), while the most recent common ancestor of the '*Cota*-group' (clade V) is reconstructed to have diverged in the central-eastern Mediterranean (HJ/IJ; Fig. 3.4a). A dispersal event into the Aegean region (I) gave rise to an ancestor located in the central-eastern Mediterranean (F/FI). From there, the clade IV spread to the Arabian peninsula (F). In the remaining part of the tree three main patterns can be recognized: (1) in clade III dispersal events to the East (IJ, GJ) are followed by a split between eastern Anatolia and western Asia (GJ → G+J) and a successive range expansion in the latter region (G), leading to the radiation of a group with the easternmost distribution of whole *Anthemis* s.l. (Fig. 3.4b);

(2) in clade II successive dispersal events followed by vicariance occurred towards the eastern (FI → I+F and DI → I+D), the central (Italy and Balkan – HI → H+I) and the western Mediterranean (Spain – AI → I+A); and (3) in clade I ('*Anthemis* s.s. group') from a most recent common ancestor set in the eastern part of the Mediterranean region (FJ/IJ), a first vicariance event between the clade situated in the eastern Anatolia–Caucasian region (clade Ie) and its sister-clade (Ia-d) is reconstructed (FJ/IJ → I/FI+J). While in the former subclade (Ie) the eastern Mediterranean range extended through dispersal events towards the Arabian Peninsula (F), in the latter one (Ia-d) several dispersal events towards the central (Greece and western Anatolia – FI, I; clade Ib-d) and finally the western (North Africa and Spain – B/AB; clade Ia) Mediterranean can be observed (Fig. 3.4b).

Eco-climatological modelling

Analyses of the climatic overlap between sister nodes/taxa allowed the recognition of 55 events of 'climatic vicariance' (no overlap of climate hypervolumes; circles in Fig. 3.4). The changes in the climatic niches of the internal nodes are visualized in the PCA plot (Fig. 3.5; factor loadings in Table 3.1). The first principal component of the PCA (PC1, explaining 47% of total variation) represents an aridity gradient, while the second principal component (PC2, accounting for 25% of total variation) describes the climate seasonality. From this information it is therefore possible to infer the climate types that characterize each panel of the PCA plot, ranging from the Mediterranean-type climate of panel I, to the humid montane-type climate of panel III. As Fig. 3.5a shows, the basal ancestors are characterized by montane

---

Next two pages: Fig. 3.4. Optimal reconstructions of geographical ancestral distribution (reconstructed through a DIVA analysis) and of climatic overlaps (reconstructed through hypervolumes exclusion) on the maximum likelihood cladogram of nrDNA ITS sequence data for *Anthemis* s.l. Symbols: circles: "climatic vicariance" events; dots: geographical vicariance events; dots in circles: climatic and geographical vicariance events. "**x**": ancestral nodes whose climatic niche is shown in Fig. 3.5a. (a): lower part of the tree. (b): upper part of the tree.

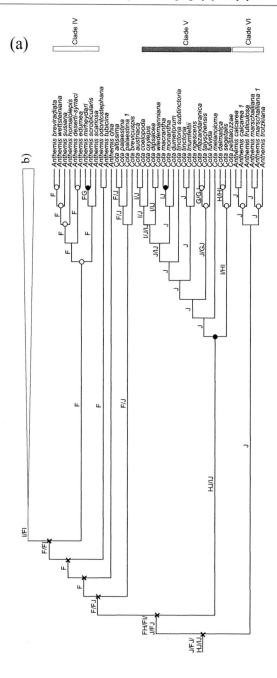

(b)

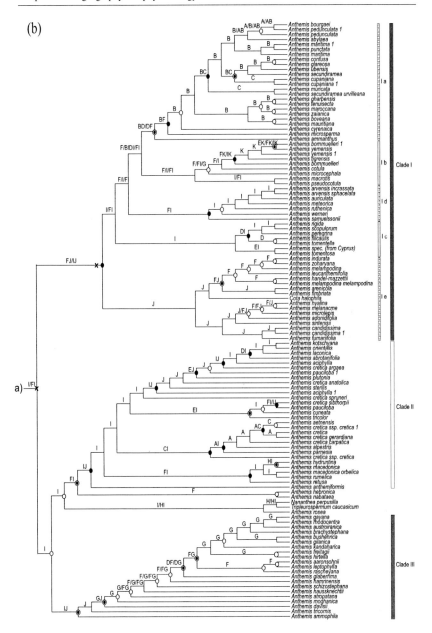

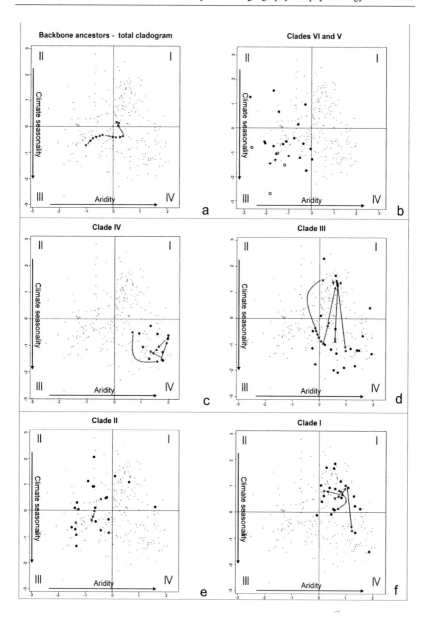

climate and a trend towards a more arid and a typically Mediterranean climate can be observed. While clade VI, one of the most basal lineages, and clade V (the '*Cota* group') show a typical montane-climate (Fig. 3.5b), clade IV is characterized by a continental-arid climate (Fig. 3.5c). The two sister clades III and II show two very different climates: while clade II is characterized by a typical montane climate (Fig. 3.5e), clade III shows a more arid climate (Fig. 3.5d). Finally, in clade I (Fig. 3.5f) a trend towards the typical Mediterranean climate is reconstructed.

Integration of temporal, biogeographical and phylo-ecological analyses

The temporal, geographical and climatological reconstructions for each clade, along with the percentages of geographical and eco-climatological speciation events, are summarized in Table 3.2. The numbers of both geographical and 'climatic vicariance' events are unevenly distributed across the tree. In clade IV, 62.5% of the nodes show no overlap in the climatic niches, a value that greatly differs from the grand mean of the 'climatic vicariance' events through the whole tree (34.4%). The same picture is shown by clades VI and III, with a percentage of 'climatic vicariance' events of 66.7% and 60%, respectively. In all these clades (IV, VI and III), the number of geographical vicariance events is lower (12.5%, 0% and 25%, respectively) than the grand mean (19.4%), suggesting only a marginal role of the geographical forces in the radiation of these groups. An opposite pattern can be identified in clade II, in which the percentage of

---

Previous page: Fig. 3.5. Principal components analysis (PCA) plots showing climatic niches at internal nodes and terminals. For each clade the climatic niches of the nodes showing a "climatic vicariance" event (circles in Fig. 3.4) are highlighted. Panels: I: Mediterranean-type climate; II Oceanic-humid-type climate; III: Humid-montane-type climate; IV: Continental-arid-type climate. The abscissa represents PC1 and shows an increasing aridity from left to right. The ordinate represents PC2 and shows decreasing climate seasonality from bottom to top. (a) Backbone ancestors – total cladogram. Symbol "×" refers to the ancestors ("×") shown in Fig. 3.4; the arrows connect nodes that are consecutive in the phylogeny (see Fig. 3.4). (b) Clades VI and V. "+" and circles refer to clade VI; "+": ancestors; circles: terminal taxa. "×" and dots refer to clade V; "×" ancestors; dots: terminal taxa. (c-f) Clades IV, III, II, and I. "×": ancestors; dots: terminal taxa; the arrows connect nodes that are consecutive in the phylogeny (see Fig. 3.4).

geographical vicariance events (37.9%) is much higher than the grand mean, and the number of 'climatic vicariance' events is very low (17.2%).

The vicariance-events-through-time plots (Fig. 3.6) allow us to recognize increases/decreases in the number of vicariance events through the different time-slices. While the percentage of geographical vicariance events remains more or less constant through time, the 'climatic vicariance' events are relatively high between 12 and 7 Ma, decrease around 7-3.5 Ma, and finally rise again in the last 3.5 Ma.

Discussion

Despite some uncertainty owing to a lack of topological support at the basal nodes of the tree, the integration of the biogeographical and the climatological analyses suggests that the ancestor of *Anthemis* s.l. was originally present in the Near East and in habitats characterized by a montane climate (Table 3.2, Fig. 3.5a). With the increase of land surfaces in the Mediterranean, which reached its maximum in the Messinian Salinity Crisis (5.96 Ma; Hsü, 1972; Krijgsman, 2002), and the progressive aridification of this region (which started around 9-8 Ma; Ivanov & al., 2002; Fortelius & al., 2006; Van Dam, 2006), a westward expansion of this plant group began. The colonization of the Mediterranean Basin in an east–west direction is a common pattern, observed both in animals (Blondel & al., 1996; Sanmartín, 2003) and in plants (Caujapè-Castells & Jansen, 2003; Lledó & al., 2005; Paun & al., 2005; Petit & al., 2005; Mràz & al., 2007). The genus *Anthemis* shows this pattern not only in the overall phylogeny, but also in clade I, the '*Anthemis* s.s. group' (Fig. 3.2b, Table 3.2). While clade Ie began to differentiate around 8-7 Ma in the eastern part of the Mediterranean, clade Ia (Fig. 3.2b) dispersed into the W Mediterranean around 7-6 Ma (Table 3.2), following the above mentioned progression of aridity towards the west.

As Hellwig (2004) pointed out, the Mediterranean region is particularly suitable for the investigation of the results of different ecological radiations: on one hand,

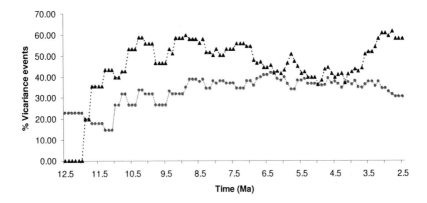

Fig. 3.6. Percentage of climatic (black, dotted line; triangles are the measured values) and geographical (grey, continuous line; dots are the measured values) vicariance events through time as inferred by adopting the sliding-window technique (time interval=5 Ma) on the dated maximum likelihood tree based on nrDNA ITS sequence data for *Anthemis* s.l.

the increasing aridity that caused the gradual disappearance of the typical sub-tropical flora (Thompson, 2005) led to the occupation of new habitats; on the other hand, the ongoing formation of mountain chains allowed the expansion towards new areas characterized by montane climate. The influence of such a changing climate in the ecological differentiation of the organisms can be found in many Mediterranean plant groups: for example, the main differentiation of *Ranunculus* L. (Ranunculaceae) into ecological groups was observed between 10 and 7 Ma (Paun & al., 2005). In general, the radiation of the genus *Anthemis* in the Mediterranean region is also accompanied by an ecological differentiation that occurred between 10 and 8 Ma (clades I to IV), as the climate in this region began to change drastically (Fig. 3.5c-f, Fig. 3.6). In particular, it is possible to observe three different patterns of ecological radiations from a common ancestor characterized by a montane-humid climate: a first lineage (clade II) dispersed into mountainous habitats in relatively recent time (from c. 2 Ma onwards), probably in connection with the glaciations of the Quaternary (Table 3.2, Fig. 3.5e). Clades IV and III dispersed into the inland regions of the Near East (into the Arabian Peninsula around 10 Ma and into the arid region of western Asia around 5 Ma, respectively); there, probably in conjunction with a progressive aridification of the climate, these two groups radiated (Table 3.2, Fig. 3.5c-d). A third lineage (clade I) evolved together with the progressive emergence of the Mediterranean climate from around 9-8 Ma onwards and moved westwards into the whole Mediterranean region (Table 3.2, Fig. 3.5f). These two latter eco-phylogenetic patterns (from wetter mountains to drier coastal plains) seem also to be a general feature for another Mediterranean region, the Cape region, where the ecological evolution within the genus *Thamnochortus* (Restionaceae) was found to have been from well-drained mountain slopes into the habitats of the coastal plain (Linder & Hardy, 2005). In the genus *Anthemis* s.l., a similar adaptation to more arid climates is also accompanied by the switch of life forms from perennial to annual, as also observed for many other Mediterranean plant groups (Cowling & al., 1996a), for example for the subtribe Centaurineae within the Asteraceae (Hellwig, 2004).

The relative frequency of geographical and climatic vicariance events varies across the clades and through time (Fig. 3.6, Table 3.2). The percentage of geographical vicariance events is particularly high in clade II, which comprises 19 perennial species distributed in the mountains of the circum-Mediterranean region. As the diversification of this clade began at the Plio-Pleistocene boundary (around 2 Ma), its radiation seems to be related to the glaciation cycles of the Quaternary, a pattern also observed for other montane plant groups (Comes & Kadereit, 1998; Kadereit & Comes, 2005; Mráz & al., 2007). Range expansions in climatically favourable periods were followed by retreats into refugia, a process known to trigger speciation through geographical vicariance and genetic isolation (Kadereit & Comes, 2005). Clade II shows a high degree of niche conservatism, detectable from its low percentage of 'climatic vicariance' events. This is consistent with the idea that geographical isolation more than adaptation to new ecological niches was the driving force for the radiation of this group.

An opposite pattern with a particularly high percentage of 'climatic vicariance' events, thus suggesting a process of niche evolution, is observed in clade III, which comprises 21 annual species widespread in western Asia (Iran, Turkmenistan, Afghanistan, Pakistan) and in clade IV, which encompasses nine annual species occurring in the northern part of the Arabian Peninsula. Although the radiation of these groups remained confined to the same geographical regions of the Near and Middle East (regions J, F, and G for clade III, and regions F and G for clade IV, respectively) and hence show low levels of geographical vicariance events (25% and 12.5%, respectively; Table 3.2), the reconstructed ancestors show high levels of niche differentiation (60% and 62.5% climatic vicariance events, respectively). This 'volatility in niche evolution' (exemplified by the vagrant positions of ancestors in clade III according to climate seasonality in the PCA of Fig. 3.5d) may be connected with the evolution of both clades in arid environments. The annual life-form, which has been promoted as one possible adaptation of plants to arid environments (Fiz & al., 2002; Datson & al., 2008), and which characterises members of clades III and IV, may have allowed faster adaptation to different environments as a consequence of shorter generation times, faster turn-over of populations, and strongly fluctuating effective population sizes, eventually leading to the observed elevated levels of climatic vicariance events.

The LTT plot shows a constant increment of lineages through time (Fig. 3.3). The onset of the Mediterranean climate (around 3 Ma; Suc, 1984; Thompson, 2005) seems to have had no effect on the diversification rate of the whole genus, a pattern also observed in a typical genus of the Mediterranean Cape Region, *Protea* (Proteaceae; Linder, 2005; Barraclough & Reeves, 2005). Conversely, other studies both for Mediterranean and alpine climates demonstrate that this is not a general trend and that climatic changes can have a strong influence on diversification rates, as exemplified by climate-triggered radiations found in *Pelargonium* (Geraniaceae) in the Cape region (Bakker & al., 2005), *Drosera* (Droseraceae) in the Australian Mediterranean region (Yesson & Culham, 2006), or various plant lineages in the European Alpine system (Kadereit & Comes, 2005).

When looking at the variation of the vicariance events through time (Fig. 3.6), three periods may be identified, in which the relative weight of geographical vs. 'climatic vicariance' events varies. Since the pattern observed is mainly caused by a considerable fluctuation in the amount of 'climatic vicariance' events through time, a plausible explanation for it may be found in the climatological history of the Mediterranean basin. The differentiation of *Anthemis* in the first period, between 12 and 7 Ma, coincides with a trend toward aridification in the Late Miocene (Ivanov & al., 2002; Fortelius & al., 2006; Van Dam, 2006). This process may have triggered a radiation of the group into habitats successively cleared from the disappearing sub-tropical flora of the Miocene (cf. Hellwig, 2004), and thus leading to the preponderance of 'climatic vicariance' events over geographical ones. The reduction of the number of climatic vicariance events in the following period, between 7 and 3.5 Ma, may be ascribed to a saturation of habitats and the return of a balance between geographical and ecological differentiation processes. The constant increment of lineages in the LTT plot (Fig. 3.3), however, argues against an overall stasis of speciation and in favour of elevated levels of diversification not explained by geography and/or eco-climatology. From 3.5 Ma onwards, the Mediterranean experienced dramatic changes in its climate, with the stabilization of the summer drought (at c. 3 Ma; Thompson, 2005; Suc, 1984) and, later, with the onset of the glacial–interglacial oscillations (at around 2.4 Ma; Bertoldi & al., 1989; Combourieu-Nebout, 1993). These events may thus explain a new increase in the number of climatic vicariance events in the last 3.5 Ma. We may however remember that the eco-climatological reconstruction has not encompassed all possible environmental variables that are involved in the definition of the niche of a species (cf. Graham & al., 2004; Yesson & Culham, 2006). A more accurate analysis could probably reveal overlaps between niches in other dimensions and, as Linder & Hardy (2005) pointed out, the circumscription of ecological niches needs to be improved.

The uniform level of the geographical vicariance events through time argues against a break in the constancy of influence of large-scale geographical modifications on the diversification of the group. The Messianian Salinity Crisis, an event regarded by many authors as a key factor for the evolution of plant diversity in the Mediterranean through allopatric speciation (e.g. Greuter, 1979; Hellwig, 2004), seems to have caused no increment in the amount of geographical vicariance events observed. However, we may consider that the geographical reconstruction through DIVA relies on relatively large predefined 'areas', and sympatric speciation events could correspond to allopatric distribution patterns on a finer geographical scale (cf. Oberprieler, 2005).

In summary, the results presented here involved phylogenetic, geographical and eco-climatological reconstructions; joint analyses of all these aspects have identified important and diverse mechanisms involved in shaping the distribution history of *Anthemis*. Despite some shortcomings of the present analyses (single molecular marker for phylogenetic reconstructions, low resolution at the basal nodes, low number of calibration points for tree dating), the here proposed joining of methods represents a tool with a considerable potential for the comprehension of distribution patterns in the Mediterranean region.

Acknowledgments

This research was supported by the Bavarian Research Foundation (BFS), by the German Academic Exchange Service (DAAD), by the German Research Foundation (grant OB 155/7-1) and by the SYNTHESYS project of the EU to RMLP and CO (AT-TAF-1618, AT-TAF-1731). We would like to thank E. Welk for his introduction in the GIS methods, S. Dötterl for his help in the GIS analyses, D. Ackerly, T. Garland and C. Yesson for their valuable assistance in computing the phylogenetic signal, G. Mansion for his help in understanding the calculation of error bars in r8s, P. Hummel for his technical support in the laboratory at the University of Regensburg and all the keepers of the Herbaria visited (B, BC, M, S, W, WU). We thank E. Conti for organizing the conference in Zurich and for inviting C.O. to give a talk on the biogeography of Compositae–Anthemideae. We also thank H.P. Linder and two anonymous reviewers whose comments improved the present manuscript considerably.

Chapter 4

Speciation in progress: phylogeography of the *Anthemis secundiramea* group across the Sicilian Channel

Materials and methods

Anthemis secundiramea group

The group of *Anthemis secundiramea* is made up of six closely related, annual, diploid taxa: the coastal central (and western) Mediterranean *A. secundiramea,* the Sicilian inland endemic *A. muricata,* the coastal *A. urvilleana* endemic to Malta, and the N African inland species *A. confusa, A. glareosa,* and *A. ubensis* (Oberprieler, 1998). Together with three other annual species, endemic to Libya (*A. kruegeriana, A. cyrenaica* and *A. taubertii*), they form the series *Secundirameae* Yavin (Yavin, 1972). According to nrDNA ITS sequence information (see chapter 2 and chapter 3), all these species are part of a well supported clade, which also includes the perennials *A. maritima, A. pedunculata* (both widespread in the west Mediterranean), *A. cupaniana* (endemic to the mountains of Sicily) and *A. abylaea* (endemic to the mountains of Morocco).

The differences among the representatives of this group are in the micro-morphology and/or related to their habitus: both *A. confusa* and *A. glareosa* are characterized by prostrate stems and tuberculate achenes, but only the former has mucronate pales and conspicuously inflated disc florets, whereas the latter shows peduncles that remain slender at maturity. *A. ubensis* and *A. secundiramea* differ from the former two taxa mainly in the more dissected leaves, whose ultimate segments are triangular to elliptic in *A. ubensis* and obovate in *A. secundiramea*. These two species differ from each other also in the habitus, which is erect in the former and prostrate in the latter. *A. urvilleana* is characterised by a more compact habit, lack of peduncles and smaller capitula, and *A. muricata* resembles in its prostrate habitus *A. secundiramea*, but lacks ligulate florets.

The reproductive biology of these species has never been studied, but successful crossing experiments (Mitsuoka & Ehrendorfer, 1972) have shown that occasional hybridization among representatives of the genus *Anthemis* is possible. Selfing experiments conducted by Uitz (1970) both on annual and perennial representatives of the tribe Anthemideae were successful only for the annuals. Pollination by unspecific insects has been reported for other representatives of the genus, e.g. *A. arvensis* (Muller, 1883), while the presence of tuberculate achenes (as in *A. secundiramea*) has been related to animal dispersal (Benedí i Gonzàlez, 1987).

Sampling strategy and DNA isolation

A total of 290 individuals from 42 populations belonging to the six species of the *Anthemis secundiramea* group were sampled in 2004-2009 (Table 4.1, Fig. 4.1). *A. secundiramea* was represented by 25 populations distributed in the Sicilian west-coast, in the islets around Sicily (Lampedusa, Pantelleria, Egadi islands, Ustica), along the south coast of France and in the N African coastland. The point-endemic *A. muricata* was included in the analysis with one population, *A. urvilleana* with four populations from Malta and one population from Gozo. Of the N African species included in the analyses, three populations belonged to *A. ubensis*, five to *A. confusa* and two to *A. glareosa*. Up to 10 individuals per population were sampled and plant material from each individual was dried in silica gel. DNA was extracted from 10-20 mg crushed leaf material according to a modified protocol based on the method of Doyle & Doyle (1987).

cpDNA amplification and sequencing

Two cpDNA regions were sequenced for four to five individuals for each of the 42 populations, for a total of 207 individuals (see Table 4.1). The two noncoding spacer regions *psbA-trnH* and *trnC-petN* were chosen, as a more comprehensive

Table 4.1. Sampling sites and sample sizes for cpDNA and AFLP analyses for the 42 populations of the *A. secundiramea* group, along with haplotypes detected in each population.

Population	Taxon	Locations and Coordinates	Collector(s), date of collection	N cpDNA	N AFLP	Haplotype
1*	*A. secundiramea* Biv.	West Coast Sicily (Italy) - 38.02N, 12.52E	Lo Presti, M., Lo Presti, R.M., 12.05.06	5	7	H01, H08
2*	*A. secundiramea* Biv.	West Coast Sicily (Italy) - 37.94N, 12.48E	Lo Presti, M., Lo Presti, R.M., 12.05.06	5	10	H01
3*	*A. secundiramea* Biv.	West Coast Sicily (Italy) - 37.86N, 12.44E	Lo Presti, M., Lo Presti, R.M., 11.05.06	5	7	H01
4*	*A. secundiramea* Biv.	West Coast Sicily (Italy) - 37.84N, 12.46E	Lo Presti, M., Lo Presti, R.M., 11.05.06	5	9	H01
5*	*A. secundiramea* Biv.	West Coast Sicily (Italy) - 37.73N, 12.48E	Lo Presti, M., Lo Presti, R.M., 09.05.06	5	8	H01
6*	*A. secundiramea* Biv.	West Coast Sicily (Italy) - 37.57N, 12.66E	Porcasi, G., Lo Presti, R.M., 06.05.04	5	7	H01, H07
7	*A. secundiramea* Biv.	Favignana, Egadi Islands (Italy) - 37.95N, 12.28E	Lo Presti, M., Lo Presti, R.M., 10.05.06	5	-	H01
8*	*A. secundiramea* Biv.	Favignana, Egadi Islands (Italy) - 37.92N, 12.28E	Lo Presti, M., Lo Presti, R.M., 09.05.06	5	8	H01
9*	*A. secundiramea* Biv.	Favignana, Egadi Islands (Italy) - 37.91N, 12.32E	Lo Presti, M., Lo Presti, R.M., 09.05.06	5	9	H01, H02
10*	*A. secundiramea* Biv.	Favignana, Egadi Islands (Italy) - 37.91N, 12.36E	Lo Presti, M., Lo Presti, R.M., 11.05.06	5	9	H01
11*	*A. secundiramea* Biv.	Favignana, Egadi Islands (Italy) - 37.93N, 12.35E	Lo Presti, M., Lo Presti, R.M., 11.05.06	5	7	H01, H02
12*	*A. secundiramea* Biv.	Levanzo, Egadi Islands (Italy) - 37.99N, 12.33E	Brancazio, L., Lo Presti, R.M., 10.05.05	5	8	H01, H05
13*	*A. secundiramea* Biv.	Levanzo, Egadi Islands (Italy) - 38N, 12.35E	Brancazio, L., Lo Presti, R.M., 10.05.05	5	10	H01
14*	*A. secundiramea* Biv.	Marettimo, Egadi Islands (Italy) - 37.99N, 12.06E	Meister, J., Himmelreich, S., Lo Presti, R.M., 22.05.06	5	7	H01, H02
15*	*A. secundiramea* Biv.	Marettimo, Egadi Islands (Italy) - 37.97N, 12.07E	Meister, J., Himmelreich, S., Lo Presti, R.M., 22.05.06	5	8	H01
16*	*A. secundiramea* Biv.	Marettimo, Egadian Islands (Italy) - 37.95N, 12.08E	Meister, J., Himmelreich, S., Lo Presti, R.M., 21.05.06	5	9	H01, H02

Table 4.1. Continued.

Popualation	Taxon	Locations and Coordinates	Collector(s), date of collection	N cpDNA	N AFLP	Haplotype
17*	*A. secundiramea* Biv.	Marettimo, Egadian Islands (Italy) - 37.96N, 12.04E	Meister, J., Himmelreich, S., Lo Presti, R.M., 22.05.06	5	8	H01
18*	*A. secundiramea* Biv.	Ustica (Italy) - 38.7N, 13.16E	Melazzo, R., Lo Presti, R.M., 23.05.04	5	7	H01
19*	*A. secundiramea* Biv.	Pantelleria (Italy) - 36.82N, 12.01E	Lo Presti, R.M., 12.04.04	5	8	H01
20*	*A. secundiramea* Biv.	Lampedusa (Italy) - 35.52N, 12.6E	Lo Presti, R.M., 02.05.05	5	9	H01, H04
21*	*A. secundiramea* Biv.	Lampedusa (Italy) - 35.52N, 12.63E	Lo Presti, M., Lo Presti, R.M., 03.05.07	5	10	H01
22*	*A. secundiramea* Biv.	Lampedusa (Italy) - 35.5N, 12.62E	Lo Presti, R.M., 02.05.05	5	7	H01, H03
23	*A. secundiramea* Biv.	CapCroisette (France) - 43.37N, 5.06E	Saatkamp A., 19.04.08	4	-	H01
24	*A. secundiramea* Biv.	El Kala (Algeria) - 36.91N, 8.47E	Véla, E. 08.05.08	5	-	H01
25	*A.muricata* (DC) Guss.	S. Caterina Villarmosa (Italy) - 37.61N, 14.11E	Carini, C., Lo Presti, R.M., 02.04.04	5	4	H06
26	*A. urvilleana* (DC) R. Fern.	Gozo (Malta) - 36.05N, 14.19E	Lo Presti, M., Lo Presti, R.M., 01.05.07	5	7	H10
27	*A. urvilleana* (DC) R. Fern.	Malta - 35.99N, 14.37E	Lo Presti, M., Lo Presti, R.M., 30.04.07	5	11	H10
28	*A. urvilleana* (DC) R. Fern.	Malta - 35.95N, 14.45E	Lo Presti, M., Lo Presti, R.M., 30.04.07	5	8	H10
29	*A. urvilleana* (DC) R. Fern.	Malta - 35.85N, 14.57E	Lo Presti, M., Lo Presti, R.M., 30.04.07	5	7	H06
30	*A. urvilleana* (DC) R. Fern.	Malta - 35.83N, 14.43E	Lo Presti, M., Lo Presti, R.M., 30.04.07	5	10	H09, H11
31**	*A. ubensis* Pomel	Cap Bizerte (Tunisia) - 37.33N, 9.87E	Lo Presti, M., Lo Presti, R.M., 27.04.07	5	9	H16, H17
32**	*A. ubensis* Pomel	Cap Bon (Tunisia) - 36.8N, 10.56E	Lo Presti, M., Lo Presti, R.M., 27.04.07	5	9	H13, H17, H19
33**	*A. ubensis* Pomel	Zaghouan (Tunisia) - 36.24N, 8.79E	Lo Presti, M., Lo Presti, R.M., 26.04.07	5	8	H21
34**	*A. ubensis* Pomel	Teboursouk (Tunisia) - 36.46N, 9.25E	Lo Presti, M., Lo Presti, R.M., 26.04.07	5	8	H21, H22, H23
35**	*A. ubensis* Pomel	Zaghouan (Tunisia) - 36.39N, 10.13E	Lo Presti, M., Lo Presti, R.M., 26.04.07	5	9	H17, H21
36**	*A. confusa* Pomel	Between Sfax and Gafsa (Tunisia) - 34.68N, 9.96E	Lo Presti, M., Lo Presti, R.M., 25.04.07	5	11	H14, H16, H17

Table 4.1. Continued.

Population	Taxon	Locations and Coordinates	Collector(s), date of collection	N cpDNA	N AFLP	Haplotype
37	*A. confusa* Pomel	Between Kariz and Tozeur (Tunisia) - 34.05N, 8.24E	Oberprieler, C. (10531), Vogt, R.(16588), Gstöttl, C., 23.03.09	5	-	H16, H17
38**	*A. confusa* Pomel	Between Gabes and Kebili (Tunisia) - 33.8N, 9.5E	Lo Presti, M., Lo Presti, R.M., 24.04.07	5	7	H12, H27, H28
39**	*A. confusa* Pomel	Between Gabes and Kebili (Tunisia) - 33.83N, 9.58E	Lo Presti, M., Lo Presti, R.M., 24.04.07	4	10	H27
40	*A. confusa* Pomel	Bir Soltane (Tunisia) - 33.24N, 9.74E	Oberprieler, C. (10452), Vogt, R. (16509), Gstöttl, C., 19.03.09	4	-	H15, H24, H25, H27
41	*A. glareosa* E. A. Durand & Barratte	Between Ben Guerdane and Tataouine (Tunisia) - 33.1N, 10.11E	Oberprieler, C. (10380), Vogt, R. (16437), Gstöttl, C., 18.03.09	5	-	H15, H21, H26
42	*A. glareosa* E. A. Durand & Barratte	Between Tataouine and Remadah (Tunisia) - 32.34N, 10.35E	Oberprieler, C. (10401), Vogt, R. (16458), Gstöttl, C., 18.03.09	5	-	H12, H18, H19, H20
Total				207	290	

* Population included in the '*A. secundiramea*' subset
** Population included in the 'N African taxa' subset

phylogenetic analysis with almost all representatives of the genus *Anthemis* included the same markers (Lo Presti & al., 2010). The chloroplast *psbA-trnH* spacer was amplified using the primers *psbA* and *trnH-(GUG)* (Hamilton, 1999), while for the amplification of the *trnC-petN* region the primers *trnC* (Demesure & al., 1995) and *petN1R* or *petN2R* (Lee and Wen, 2004) were used. PCR amplifications were performed with 0.2 μM dNTP's, 0.02 μM of each primer, 0.2 U Taq polymerase (Qbiogene) in 10 μl 1x Buffer and the following temperature profile: 2-5 min at 95 °C, then 36 cycles of 30 s at 95 °C, 60 sec at 50 °C (for *psbA-trnH*) or 63 °C (for *trnC-petN*), 60 s at 72 °C, with a final extension of 5 min at 72 °C. The polymerase chain reaction (PCR) products were purified with Agencourt AMPure magnetic beads (Agencourt Bioscience Corporation, Beverly, Massachusetts, USA) and cycle sequencing reactions were performed using the DTCS Sequencing kit (Beckman Coulter, Fullerton, California, USA), following the manufacturer's manual. The fragments were separated on a CEQ8000 sequencer (Beckman Coulter, Fullerton, California, USA). Sequences obtained were aligned separately using Clustal W (Thompson & al., 1994) as implemented in BioEdit version 7.05.2 (Hall, 1999) and each alignment was optimized manually. An equivocal region in the alignment of the *trnC-petN* spacer consisting of 31 bp was excluded from all subsequent analyses. Gaps were coded as binary characters using the simple gap coding method of Simmons and Ochoterena (2000) implemented in GapCoder (Young and Healy, 2003).

To test the phylogenetic relationships between the haplotypes of the sampled populations and other Mediterranean taxa of *Anthemis*, the sequences generated from the two cpDNA markers were combined with those already available from former published accessions for 34 Mediterranean representatives of the genus *Anthemis* and 4 outgroup taxa belonging to the closely related genus *Cota* (see Appendix A1).

AFLP analysis

The Amplified Fragment Length Polymorphism (AFLP) fingerprinting was performed for 35 populations with 4-10 individuals per population (290 individuals; see Table 4.1). The procedure adopted followed the protocol of Meister & al. (2006) with minor modifications. DNA concentration was measured with the spectrophotometer NanoDrop ND-1000 (Peqlab, Erlangen, Germany) and 50 ng of genomic DNA were used for restriction and ligation reactions, which were conducted in a thermal cycler for 2 h at 37 °C with *Mse*I and *Eco*RI restriction enzymes (Fermentas, St. Leon-Rot, Germany) and T4 DNA Ligase (Fermentas, St. Leon-Rot, Germany). The following PCRs were run in a reaction volume of 5 µl and the reaction parameters followed Meister & al. (2006). Preselective amplifications were performed using primer pairs with a single selective nucleotide, *Mse*I-C and *Eco*RI-A, 0.2 µM dNTP's and 0.25U/µl Taq-Polymerase (Peqlab) in 5 µl 1x Buffer. After performing an initial screening of 27 combinations of selective primers on 4 individuals from different geographical areas, the three selected combinations were: *Mse*I-CAA/*Eco*RI-AGC, *Mse*I-CAG/*Eco*RI-AAG and *Mse*I-CAA/*Eco*RI-ACT. After DNA precipitation, DNA pellets were vacuum dried and dissolved with a mixture Sample Loading Solution (Beckman Coulter, Krefeld, Germany) and CEQ Size Standard 600 (Beckman Coulter, Krefeld, Germany). AFLP products were separated via capillary gel electrophoresis on an automated sequencer (CEQ 8000, Beckman Coulter, Krefeld, Germany). Raw data were exported to GelCompar vers. 5.1 (Applied Maths, Austin, Texas, USA) and errors in the standard were adjusted manually. AFLP fragments were automatically scored following the automatic test procedure of Holland & al. (2008). A total of 60 character matrices were generated for 43 samples, representatives of all populations, by modifying the following scoring parameters in GelCompar: minProfiling (between 0.5 and 2.5), minArea (between 0.0 and 0.5), shoulder sensitivity (between 0 and 1) and matching tolerance (between 0.02 and 0.08). For each character matrix, the resolution score and the normalized resolution score, the Euclidian and the Jaccard error rates, and the number of correctly paired replicates were recorded. The set of parameters that minimize the error rate were chosen, which are: minProfiling=1.5; minArea=0.5; shoulder sensitivity=0; matching tolerance=0.02.

For ten individuals belonging to ten different populations, the whole AFLP procedure was repeated twice, in order to evaluate the quality of the fingerprints. The error rate was calculated as the number of phenotypic differences among replicates related to the total number of bands.

Data Analyses

cpDNA - For the 207 individuals belonging to the 42 populations, the two partitions *psbA-trnH* and *trnC-petN*, consisting of a stretch of nucleotides and the associated gaps, were analysed together to construct a network of haplotypes with the software TCS vers. 1.21 (Clement & al., 2000). TCS creates an unrooted haplotype network by implementing the statistical parsimony algorithm described by Templeton & al. (1992).

The haplotypes found were united with the 34 sequences of Mediterranean representatives of *Anthemis* and the 4 outgroup taxa and a Bayesian inference (BI) phylogenetic analysis was performed with MrBayes version 3.1.2 (Ronquist & Huelsenbeck, 2003). The BI analysis implemented a GTR model with an invariate gamma distribution ('Lset NSt=6 rates=invgamma'), without fixing rates and nucleotide frequencies, as these parameters are estimated from the data during the analysis. Substitution models and rates of substitution were allowed to vary among the parameters ('unlink' command and 'ratepr=variable') and a binary model ('Lset coding=variable') was applied to the coded gaps (Ronquist & al., 2005). The analyses were conducted using three heated chains and one cold chain, with a chain heating parameter value of 0.2. The MCMC chains were run for 5,000,000 generations, with trees sampled every 1,000th generation. Convergence was checked by examining the average standard deviation of split frequencies and by comparing likelihood values and parameter estimates from different runs in Tracer v. 1.3 (Rambaut and Drummond, 2003). A 50% majority rule consensus tree of the trees sampled after the achieving of convergence was computed.

Finally, to test independently the genetic variation in a mainland system vs. a system of islands, two subsets of data were derived: the first one included 8 Tunisian populations (69 individuals) of *A. confusa* and *A. ubensis* ('N African taxa' subset) the second one comprised 21 populations (168 individuals) of *A. secundiramea* widespread in Sicily and surrounding islands ('*A. secundiramea*' subset; see Table 4.1). Genetic differentiation among populations without (G_{ST}) and with (N_{ST}) taking into account the similarities among haplotypes was estimated with Permut v. 1.0 (Pons & Petit, 1996), for each subset independently, performing 10,000 permutations.

AFLP - To detect genetically similar groups of individuals among the 35 populations, we constructed a dendrogram in PAUP* vers. 4.0b10 (Swofford,

2002) using the neighbour-joining method (NJ; Saitou & Nei, 1987) in conjunction with pairwise Nei & Li (1979) genetic distances. Support for each node was computed by bootstrapping (BS) with 10,000 replicates. A pairwise Φ_{PT} matrix among all 35 populations and based on the squared Euclidean distances among AFLP phenotypes was computed with Arlequin vers. 3.11. (Excoffier & al., 2005) and imported into GenAlEx vers. 6.2 (Peakall & Smouse, 2006) to perform a principal coordinate analysis (PCoA). For each population the genetic diversity was estimated, by measuring the Shannon index (Shannon & Weaver, 1949) in GenAlEx vers. 6.2 (Peakall & Smouse, 2006).

For each subset derived from the original presence/absence matrix for the 35 populations (subset 'N African taxa' and subset '*A. secundiramea*'; see Table 4.1), a NJ dendrogram and a PCoA were computed with the same settings described above. Moreover, genetic differentiation among populations (Φ_{PT}) was estimated in GenAlEx vers. 6.2 (Peakall & Smouse, 2006). Correlations between genetic (pairwise Φ_{PT} values among populations) and geographical distances were tested for each subset with a Mantel test (Mantel, 1967) in GenAlEx vers. 6.2 (Peakall & Smouse, 2006), using 9,999 permutations. This analysis was done to test whether the accumulation of genetic differences among populations occurred under a geographically restricted dispersal process (isolation-by-distance; Wright, 1943).

Species distribution modelling

The potential distributions of the six species included in the *A. secundiramea* group according to climatic factors were modelled using Maxent version 3.0.6 (Maximum Entropy Species Distribution Modelling; Phillips & al., 2006). In all grid cells of the study area, Maxent searches for the probability distribution that better describes the potential geographical range of each species. The probability distribution chosen is the one of maximum entropy (i.e. closest to having equal probabilities of occurrence in all grid cells), with the constraint that the expected value of each climatic variable matches the average of the variable's values of the real distribution points (Kozak & Wiens, 2006). This method seems to perform better compared to other methods such as GARP (Genetic Algorithm for Rule-set Prediction, Stockwell & Peters, 1999) or Bioclim (Nix, 1986) and it does not require specification of absences of species, allowing making inference on the potential distribution of poorly known species (Phillips & al., 2006; Kozak & Wiens, 2006; Elith & al., 2006).

The distribution points for each of the six species came from Lo Presti & Oberprieler (2009) with the addition of the 42 new locations of populations collected for the present study. They included 156 occurrence data (76 for *A. secundiramea*, 6 for *A. urvilleana*, 4 for *A. muricata*, 27 for *A. ubensis*, 14 for *A. glareosa*, and 29 for *A. confusa*) and covered the entire distribution range of the taxa. The data for current climatic conditions were obtained from the WorldClim dataset (www.worldclim.org), which contains records for the 1960-90 period with a 2.5 min (approximately 5 km^2) resolution (Hijmans & al., 2005). This dataset includes 19 'bioclimatic variables', i.e. biologically meaningful variables derived from the monthly temperature and rainfall values (variables are listed in Table 3.1). Finally, the overlapping-percentages among the predicted distributions of all species were calculated.

Results

Chloroplast variation and geographical distribution of haplotypes

The length of the two cpDNA regions for 207 individuals ranged between 820 and 852 bp, and resulted in an alignment of 880 bp (896 characters with indels coded). In the *psbA-trnH* region, 22 polymorphisms were detected across the whole dataset, of which 11 were indels and 11 were substitutions, while in the *trnC-petN* spacer only 5 indels and 9 substitutions (14 polymorphisms) were found. The number of single nucleotide polymorphisms (SNP) however, was higher for *trnC-petN* (where 12 SNP were identified) than for *psbA-trnH* (7 SNP). All mutations together defined 28 haplotypes, of which 14 were singletons and 5 were shared by up to three individuals (see Fig. 4.1).

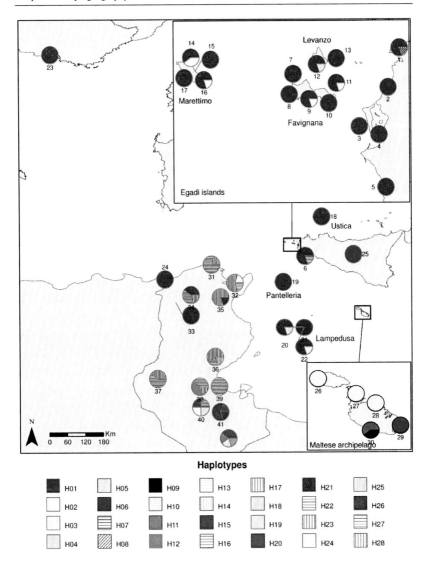

Haplotypes

■ H01	☐ H05	■ H09	☐ H13	▥ H17	■ H21	☐ H25
☐ H02	■ H06	☐ H10	▦ H14	☐ H18	☰ H22	■ H26
☐ H03	☰ H07	▨ H11	☰ H16	☐ H19	▦ H23	☰ H27
☐ H04	▨ H08	▨ H12	☰ H16	■ H20	☐ H24	▦ H28

Fig. 4.1. Geographical distribution and frequencies of the 28 haplotypes detected in the 42 populations of the *A. secundiramea* group. The relationships among haplotypes are shown in Fig. 4.2. Pops. 1-24: *A. secundiramea*. Pop. 25: *A. muricata*. Pops. 26-30: *A. urvilleana*. Pops. 31-35: *A. ubensis*. Pops. 36-40: *A. confusa*. Pops. 41-42: *A. glareosa*.

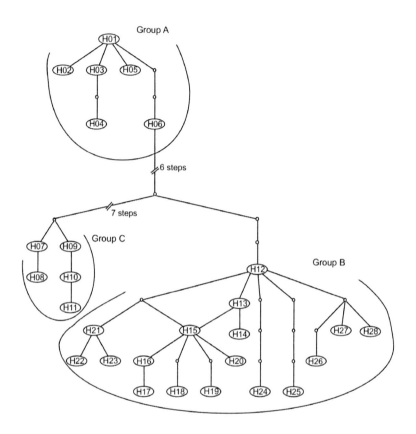

Fig. 4.2. cpDNA haplotype network obtained for 207 individuals from 42 populations of the *A. secundiramea* group. Small white circles represent intermediate haplotypes that were not detected.

TCS calculated a 95% parsimony connection limit of 13 steps and resulted in a network with three main groups, separated from each other by at least 10 steps (Fig. 4.2). The first group (*group A*), which includes 127 individuals and 6 haplotypes, comprises all populations belonging to *A. secundiramea* (except two individuals, see below), *A. muricata* and one population of *A. urvilleana* in Malta. While 4 haplotypes are singletons or shared by not more than 5 individuals, haplotype H06 is found both in *A. muricata* and in *A. urvilleana* (Fig. 4.1). H01 is recorded for 110 individuals (all belonging to *A. secundiramea*) and, representing 53% of all samples, is the most common haplotype and has the largest geographical distribution (Fig. 4.1). The second group (*group B*) includes 17 haplotypes harboured in 58 individuals belonging to the 12 populations of *A. ubensis, A. confusa* and *A. glareosa* located in Tunisia. Five haplotypes are recorded across taxa: H12 and H15 are found in populations belonging to *A. confusa* and *A. glareosa*, H21 is shared by *A. glareosa* and *A. ubensis*, while H17 occurs in individuals of *A. confusa* and *A. ubensis*.

Three of the five haplotypes included in the third group (*group C*) are found in the remaining four populations (20 individuals) of *A. urvilleana* in Malta and Gozo, while the last two (H07 and H08) are singletons recorded for two individuals of *A. secundiramea*. The three groups identified by TCS were also found in the BI tree with other Mediterranean representatives of *Anthemis* (Fig. 4.3). *Group C*, which also includes chloroplast haplotypes of the two annual species *A. peregrina* and *A. arvensis*, is well supported and is basal to the other two groups. *Group A* is part of a well supported clade that comprises other annual species distributed in eastern Mediterranean, while all haplotypes forming *group B* are included in a well supported clade together with other taxa, both annual and perennial (e.g. *A. pedunculata*), widespread in northern Africa.

Table 4.2. Mean estimates of haplotype diversity among populations of the '*A. secundiramea*' and the 'N African taxa' subsets. G_{ST} and N_{ST} are based on cpDNA data and were estimated with Permut vers. 1.0 (Pons & Petit, 1996). Φ_{PT} is based on AFLP data and was estimated with GenAlEx vers. 6.2 (Peakall & Smouse, 2006).

Subset	'*A. secundiramea*'	'N African taxa'
Number of populations	21	8
G_{ST}	0.018	0.469
N_{ST}	0.031	0.703
Φ_{PT}	0.086*	0.054*

* P = 0.01

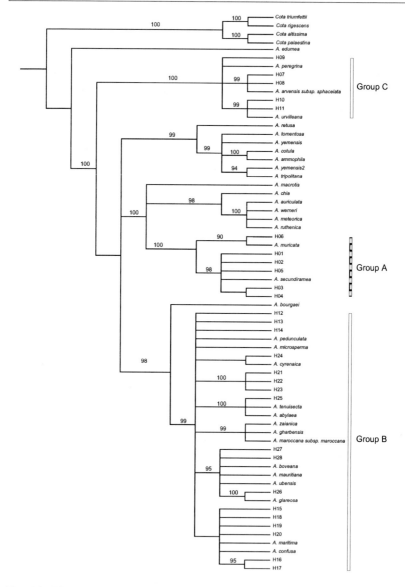

Fig. 4.3. BI tree of cpDNA data (*psbA-trnH* and *trnC-petN* spacer regions) for 34 Mediterranean representatives of *Anthemis*, 4 outgroup taxa (*Cota* spp.) and 28 haplotypes detected for the *A. secundiramea* group. Numbers above the branches indicate support values (only PP ≥ 0.95 are shown).

Measures of genetic diversity showed that among the N African populations N_{ST} was significantly higher than G_{ST}, thus demonstrating the existence of a phylogeographic structure, as closely related haplotypes are more likely to be found in the same rather than different populations. On the other hand, the populations of *A. secundiramea* lacked such a structure, as N_{ST} was not significantly higher than G_{ST} (Table 4.2).

Population structure based on AFLP

The three primer combinations applied to 290 individuals belonging to 35 populations generated 1,297 fragments, ranging from 100 to 500 bp. (D2 with 476, D3 with 449 and D4 with 372 fragments, respectively). An average error rate of 7.36% was estimated among all replicates. The NJ analysis conducted on all individuals revealed a subdivision into three groups (Fig. 4.4a). Group I comprises *A. secundiramea* and a subclade representing *A. muricata*, while group II encompasses all individuals belonging to the two Tunisian species *A. ubensis* and *A. confusa*. Group III includes all individuals belonging to *A. urvilleana*. Within each of the three groups, any of the subclades identifies distinct populations (with the above exception of *A. muricata*). The same groups are revelead also by the PCoA of all populations (Fig. 4.4b), with the only exception of *A. muricata*, which clusters closer to the Tunisian group (group II) than to *A. secundiramea* (group I). The first and the second axes account for 35.4% and 21.7% of the total variance, respectively. The Shannon Indexes for all populations (Fig. 4.5) show that the genetic diversities are similar in all *A. secundiramea* populations and range between 0.110 and 0.148, while the N African populations are characterized by higher values. *A. muricata* (pop. 25, Fig. 4.1) and one of the Maltese populations (pop. 29, Fig. 4.1) show the lowest values of genetic diversity (0.074 and 0.097, respectively).

The analyses carried out for all individuals widespread in the W coast of Sicily and surrounding islands and assigned to *A. secundiramea* (21 populations, 168 individuals: see Table 4.1) revealed the absence of population structure. In the NJ dendrogram (Fig. 4.6a), individuals cluster together independently from their affiliation to a population, while in the PCoA (Fig. 4.6b; first axis accounting for 33.5%, second axis for 25.1% of the total variance) the populations located in the same island do not cluster together. The only exception is represented by Lampedusa: The three populations located on this southernmost island belonging to Italy are distinctly separated from all other populations in the PCoA. The absence of a pattern of isolation by distance is confirmed by the Mantel test, which showed no significant correlation between geographical distances and pairwise Φ_{PT} values among populations ($r = 0.131$, $P = 0.18$). The total genetic differentiation among populations was 0.086 (Table 4.2).

A different scenario was observed for the 8 populations located in Tunisia. In the NJ dendrogram (Fig. 4.7a) almost all individuals belonging to *A. confusa* are clustering together. In the PCoA, the populations of *A. confusa* (except one, see Fig. 4.7b) and *A. ubensis* are well isolated along the first axis (which accounts for 37.1% of the total variance), while the populations belonging to *A. ubensis* form two well-separated groups along the second axis (explaining 22.1% of the total variance). It is interesting to note that one of these groups includes two populations (pops. 31 and 32, Fig. 4.1) that resemble in their morphology *A. secundiramea*. The Mantel test revealed a significant pattern of isolation by distance ($r = 0.660$, $P = 0.01$), showing increasing Φ_{PT} values with geographical distance, while the total genetic differentiation among populations was 0.054 (Table 4.2).

Species distribution modelling

The modelled distributions of each taxon are given in Figs. 4.8-4.10. The model of *A. secundiramea*, which covers coastal areas of Sicily, Sardinia, Corse and all minor islands in western Mediterranean, includes the envelope of *A. urvilleana*. However, the model of the latter taxon does not find any other similar area within the Western Mediterranean (with the exception of Zembra, a Tunisian island), thus indicating that the Maltese archipelago is characterized by different environmental conditions than those characterizing the niche of *A. secundiramea*. Moreover, while *A. secundiramea* could occur with excellent

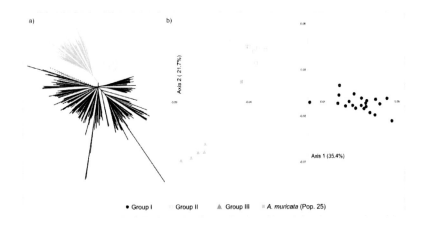

Fig. 4.4. Results of the AFLP analyses for 290 individuals belonging to 35 populations of the *A. secundiramea* group. a) NJ dendrogram with 290 individuals. b) Principal Coordinate Analysis (PCoA) of the 35 populations.

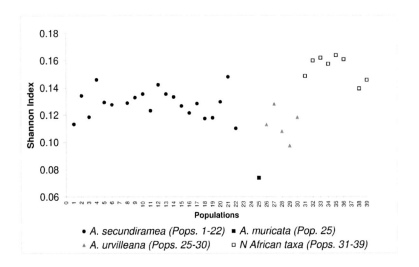

Fig. 4.5. Shannon indexes for 35 populations of the *A. secundiramea* group, as estimated by GenAlEx vers. 6.2 (Peakall & Smouse, 2006).

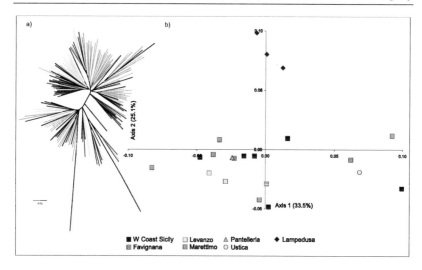

Fig. 4.6. Results of the AFLP analyses for the '*A. secundiramea*' subset (168 individuals belonging to 21 populations). a) NJ dendrogram with 168 individuals. b) Principal Coordinate Analysis (PCoA) of the 21 populations.

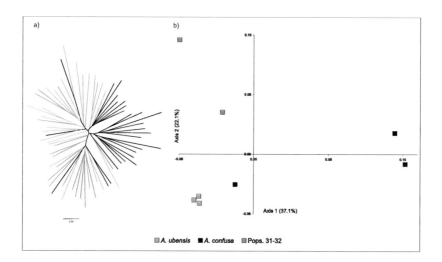

Fig. 4.7. Results of the AFLP analyses for the 'N African taxa' subset (69 individuals belonging to 8 populations). a) NJ dendrogram with 69 individuals. b) Principal Coordinate Analysis (PCoA) of the 8 populations.

probability (90-100%) in N Tunisia and in N Algeria, neither of the N African species would find optimal climatic conditions outside of Africa (with the only exception of the Maltese archipelago for *A. glareosa*). The models of the three N African species are quite different from each other: while *A. ubensis* is predicted for the mountainous regions of N Tunisia, the model of *A. glareosa* covers the southern coastal regions of Tunisia, reaching also the Maltese archipelago, and *A. confusa* is predicted for the central part of Tunisia and northern Algeria. The percentages of overlapping among the predicted distributions of all species are given in Table 4.3.

Table 4.3. Niche overlaps among the taxa belonging to the *A. secundiramea* group. Percentages of overlap are given as a fraction of the modelled areas of the species in columns. The modelled areas do not include the whole range of probability of occurrence (shown in Figs. 4.8-4.10), but are defined as the sum of grid cells with probability of occurrence higher than 75%.

	A. ubensis	*A. confusa*	*A. glareosa*	*A. secundiramea*	*A. urvilleana*	*A. muricata*
A. ubensis	**100**	0.04	0	5.38	0	0
A. confusa	0.08	**100**	60.82	0	0	0
A. glareosa	0	20.56	**100**	0.77	100	0.87
A. secundiramea	8.35	0	1.77	**100**	58	38.21
A. urvilleana	0	0	3.05	0.77	**100**	0.87
A. muricata	0	0	0.24	4.66	8	**100**

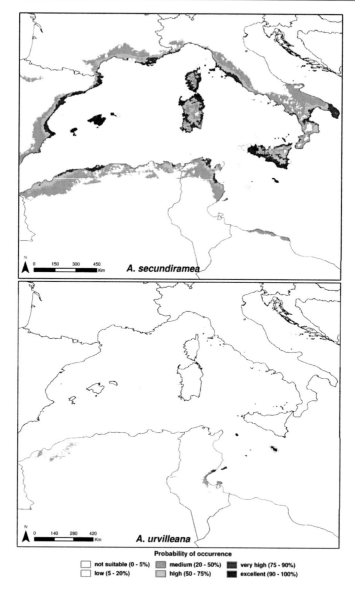

Fig. 4.8. Potential distribution of *A. secundiramea* (above) and *A. urvilleana* (below) according to climatic factors modelled with Maxent vers. 3.0.6 (Phillips & al., 2006).

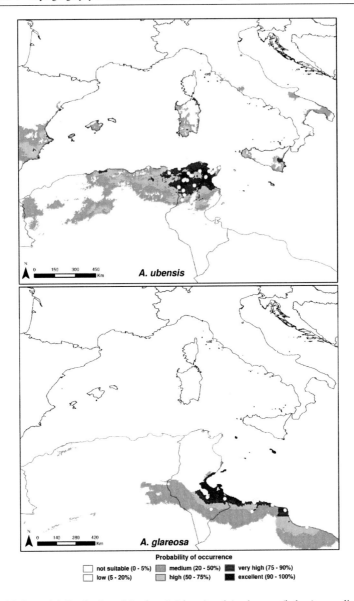

Fig. 4.9. Potential distribution of *A. ubensis* (above) and *A. glareosa* (below) according to climatic factors modelled with Maxent vers. 3.0.6 (Phillips & al., 2006).

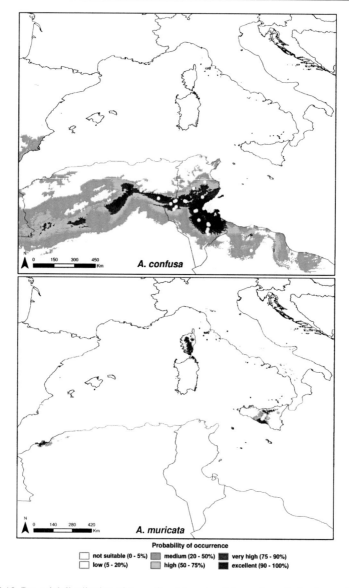

Fig. 4.10. Potential distribution of *A. confusa* (above) and *A. muricata* (below) according to climatic factors modelled with Maxent vers. 3.0.6 (Phillips & al., 2006).

Discussion

Taxonomical issues

The actual circumscription of *Anthemis* ser. *Secundirameae* Yavin consists of nine species distributed in the C and E Mediterranean area (Oberprieler, 1998). However, cpDNA analyses have shown that this series is not monophyletic: while the N African representatives of the series cluster with high support with other N African taxa, *A. muricata* and *A. secundiramea* form the sister clade of annual species widespread in the E Mediterranean (Fig. 4.3).

A. secundiramea, in particular, is the most widespread species of the whole series, and its high morphological variability, especially in Sicily and its adiacent islets, has resulted in the description of several infraspecific taxa. *A. urvilleana* has been alternatively treated at sub-specifical rank as *A. secundiramea* subsp. *urvilleana* (see Greuter & al., 2003). However, the overall genetic similarity among all populations belonging to *A. secundiramea* and, at the same time, their differences to *A. urvilleana*, both at cpDNA and AFLP level (Fig. 4.1, Fig. 4.4a, and Fig. 4.5) seem to confirm the recognition of *A. urvilleana* as separate species.

A. secundiramea, N African populations and the Sicilian Channel

Both the cpDNA and the AFLP analyses have identified two well-defined groups of taxa: one that includes all populations belonging to *A. secundiramea*, and another with the species widespread in N Africa (Fig. 4.3 and Fig. 4.4). However, some unexpected patterns were revealed. *a)* Two Tunisian populations (pops. 31-32, Fig. 4.1), collected in the localities reported in literature for *A. secundiramea* (Oberprieler, 1998, and references therein: Cap Bon and Cap Blanc) and which were morphologically assigned to this species whilst sampling, revealed a stronger connection to the Tunisian species *A. ubensis* and *A. confusa*, both in the chloroplast and in the AFLP-fingerprint analyses (Fig. 4.1 and Fig. 4.4a). Only when analysing the N African populations separately, the two populations revealed a quite isolated position in the PCoA from both taxa (Fig. 4.7b). *b)* One population located in Algeria (pop. 24, Fig. 4.1) shares the same haplotype with *A. secundiramea. c)* The populations sampled in the island of Lampedusa (pops. 20-22, Fig. 4.1), which is located on the continental shelf of Africa and was connected to Tunisia during the glacial periods of the Quaternary (Giraudi, 2004), show no similarity with the N African taxa, and are included in the group of *A. secundiramea* (Fig. 4.1 and Fig. 4.4). Thus, while no one of the N African taxa is found outside of Africa, *A. secundiramea* reaches Algeria and the island of Lampedusa, but not Tunisia.

Dating information obtained from a ITS (Lo Presti & Oberprieler, 2009; see also chapter 3) and a combined ITS-cpDNA datasets (chapter 2) has shown that the African clade is c. 5 Ma old, an age that roughly matches the end of the Messinian Salinity Crisis and the reinundation of the Mediterranean (5.33 Ma; Hsü, 1972, Krijgsman, 2002). We may, therefore, suppose that this event gave rise to the ancestor of the African clade through a process of geographical vicariance. Given the low mutation rates that characterize the evolution of cpDNA (Wolfe & al., 1987; Provan & al., 1999), also the high genetic distance of this group from the other two (see Fig. 4.2) suggests an old differentiation (Ronikier & al., 2008). *A. secundiramea*, on the other hand, has originated much later, at about 1.6 Ma ago (Lo Presti & Oberprieler, 2009; see also chapter 3), after the definitive separation between Africa and Europe. After the desiccation of the Mediterranean in the Miocene, Africa and Europe have never been connected, as the depression between the two continental shelfs reaches abyssal depths of more than 3,000 m (Cassar & al., 2008; Giraudi, 2004; Wallmann & al., 1988) and the Quaternary sea-level changes related to the glaciations have occurred in a range of hundreds of meters (Lambeck & al., 2002; Yokoyama & al., 2000). The presence of *A. secundiramea* in the N African region can be therefore explained only by successful crossing of the Sicilian Channel, which in turn might occur: (a) through stepping-stone islands, which could have been revealed during the glacial periods, as the sea level sank (Ortiz & al., 2009; Stöck & al., 2008; Collina-Girard, 2001); or (b) via occasional long-distance dispersal, whose fundamental role in the explanation of biogeographic patterns is increasingly being recognized (Cowie & Hoolland, 2006) and has often been invoked in the Mediterranean (e.g., to explain the disjunct distribution of *Armeria pungens* in Portugal and Sardinia/Corse; Piñeiro & al., 2007).

The presence of *A. secundiramea* in areas that have never been connected to each other is not limited to Algeria or the island of Lampedusa. This taxon has successfully colonized other Mediterranean islands such as Ustica or Pantelleria, which are of volcanic origin and since their emergence (at about 500 ka and 114 ka, respectively; de Vita & al., 1998; Wallmann & al., 1988) have never been connected to Sicily or Tunisia; or the Balearic islands, which are of continental origin but have remained isolated since the opening of the Gibraltar Strait (about 5.3 Ma; Gautier & al., 1994), thus long before the appearance of *A. secundiramea*; and finally Corse, whose even more ancient isolation (about 16 Ma; Speranza & al., 2002) was partially interrupted only during the Messinian Salinity Crisis in the Miocene (c. 5 Ma; Hsü, 1972; Duggen & al., 2003; Mansion & al., 2009). Hence, even if it has not yet been studied, the dispersal ability of *A. secundiramea* seems to be excellent.

Therefore, given that the absence from Tunisia is not an artefact derived from incomplete sampling, the issue remains unresolved why *A. secundiramea* is not found in Tunisia, as also the modelled potential distribution shows that this taxon could occur with very high to excellent probability (Fig. 4.8) along its coasts. Two scenarios seem to be plausible: 1) *A. secundiramea* has never crossed the Sicilian Channel and its casual presence in the N African region is associated to a human-mediated dispersal, a very common event in the Mediterranean, where the activity of man has a long history (Blondel, 2006; Thompson, 2005; Martinez & Montero, 2004) and has shaped the genetic patterns of many taxa (e.g., *Quercus*; Petit & al., 2002; or *Pinus pinea*; Vendramin & al., 2008). 2) *A. secundiramea* has reached the N African region, but a successful colonization of the Tunisian coasts by sporadic individuals was circumvented by the competition with the well-established local populations. As it has been stressed that the region of Cap Bon (Tunisia), but not W Algeria (where *A. secundiramea* is found) has been a glacial refugial area (Médail & Diadema, 2009), it is plausible that a dispersal of *A. secundiramea* into the former region may have not been successful at all, or has been followed by hybridization and strong introgression, which could explain the quite isolated position of the two Tunisian populations from the other N African taxa in the PCoA (Fig. 4.7). The position of *A. secundiramea* among the N African taxa only in the ITS BI tree (Lo Presti & Oberprieler, 2009; see also chapter 3) and not in the analyses of the complete genome (AFLP; Fig. 4.4) may further support this hypothesis.

A. *urvilleana* in the Maltese archipelago and its connections to Sicily

The AFLP fingerprint analyses have recognized a third group (group III, Fig. 4.4) that includes all populations belonging to *A. urvilleana* located in the Maltese archipelago. The four haplotypes found in these populations, however, are included in two separate groups, both in the haplotype network and in the BI tree (group A and group C, Fig. 4.2 and Fig. 4.3). The first group (group C, Fig. 4.2) comprises three haplotypes found in four populations in Malta and Gozo (pops. 26-28 and 30, Fig. 4.1) that cluster with two other genotypes found in two Sicilian populations of *A. secundiramea* (Pops. 1 and 6, see Fig. 4.1). The BI tree (Fig. 4.3) shows that they are part of a well-supported clade that comprises the cpDNA sequences of the two annual species *A. arvensis* and *A. peregrina*. As both latter species are widespread in Malta and in Sicily (see Geôrgiou, 1990), the hypothesis of hybridization/introgression among these taxa seems to be plausible. Support for the occurrence of reticulate evolution was already provided by crossing experiments both at infra- and inter-generic level within the tribe Anthemideae (among *Anthemis*, *Cota* and *Tripleurospermum*; Mitsuoka & Ehrendorfer, 1972). Therefore, occasional hybridization is likely to occur (Yavin, 1972; Grant, 1971; Stebbins, 1942).

The remaining haplotype (H06) found in only one population in Malta (pop. 29, Fig. 4.1) is included in the same group with *A. secundiramea* (group A, Fig. 4.2) and is shared with the Sicilian endemic *A. muricata*. The affinities between the Maltese and the Sicilian flora and fauna have been well documented (see Junikka & al., 2006 for the flora and Schembri, 2003 for the fauna), and are to be ascribed to the old connections among the two islands. South-eastern Sicily

(Hyblean plateau) and the Maltese archipelago (Malta plateau) are linked by a submarine ridge composed by sedimentary rocks of Tertiary age (Cassar & al., 2008), which emerged between the middle Pliocene and the Early Pleistocene (Gardiner & al., 1995; Catalano & al., 1995). This ridge is nowhere deeper than 200 m (Cassar & al., 2008; Pedley & al., 1976) and was an epicontinental land bridge during the Pleistocene (Alexander, 1988). Thus, since their emergence, the Maltese archipelago and South-Eastern Sicily have always been connected, while periods of isolations are mainly related to episodes of marine transgression during the interglacials (Cassar & al., 2008). We may therefore assume that the ultimate isolation between the two islands in the present interglacial gave origin to the two taxa through geographical vicariance. In this scenario, the shared haplotype represents a remnant of the ancient connection between the two taxa and the overall low genetic diversity that characterizes both *A. muricata* and *A. urvilleana* compared to *A. secundiramea* (Fig. 4.5) provides further insight that they are relics of a past range fragmentation connected with bottle-neck effects. However, only one of the five sampled populations of *A. urvilleana* in the Maltese archipelago still includes the ancient chloroplast type: Given that our sampling reflects the real distribution of haplotypes, it seems to be that the hybridization with *A. arvensis* or *A. peregrina* is leading to the extinction of the 'original' *A. urvilleana*. The low genetic diversity that characterizes the population with the ancient haplotype in respect to the hybrid populations (Fig. 4.5) is correlated to a high danger of extinction, whose rate is predicted to be higher for endemic than for nonendemic insular populations (Frankham, 1997).

The haplotype shared by *A. muricata* and *A. urvilleana* (H06) has an internal position relative to the typical *A. secundiramea* haplotypes (H01-H05; Fig. 4.2), thus arguing for an older origin of the former than of the latter. Moreover, as this ancient haplotype is shared by two taxa, it is more likely that the speciation event that gave rise to them is quite recent, and successive to the origin of *A. secundiramea*.

A. secundiramea in a system of islands

The populations belonging to *A. secundiramea* are widespread in regions with very different geological characteristics. On one hand, the system consisting of the Egadi islands and the W Coast of Sicily represent remnants of an old submarine orogenic belt linking the Apennine-Maghrebian and the Sicilian mountain chains (Martini & al., 2007). The whole region emerged at the end of the lower Pleistocene (c. 800 ka; Agnesi & al., 1993) and not until the present interglacial the Egadi islands have been isolated from the mainland. On the other hand, there are two volcanic islands, Ustica and Pantelleria, which since their emergence (at about 500 ka and 114 ka, respectively; de Vita & al., 1998; Wallmann & al., 1988) have never been connected to Sicily or Tunisia. Finally, the island of Lampedusa is a part of the African shelf that emerged at the Plio-Pleistocene boundary (c. 1.8 Ma; Catalano & al., 1995) and, while it has always been separated from the European continent, during the Quaternary glaciations was connected to N Africa (Giraudi, 2004).

In such a variegated landscape, however, the pattern of genetic variation is quite astonishing: there is not a significant pattern of isolation by distance and the diversity indexes of each population are quite similar (Fig. 4.5). Only a Φ_{PT} of 0.086 in the AFLP data is a signal of a moderate level of genetic structure, which however is not confirmed by its counterpart for the cpDNA data (N_{ST} does not significantly differ from G_{ST}; Table 4.2). These results could suggest that the population group is panmictic, a situation that may be related to an actual, strong gene flow among all populations. In this case, however, it remains difficult to explain: 1) the presence of local haplotypes both in the Egadi Islands and in the populations located in Lampedusa, the latter being moreover isolated from all other populations in the PCoA (Fig. 4.6); and 2) the slightly lower genetic diversity in Ustica and Pantelleria (Fig. 4.5). If we therefore assume that we do have restricted gene flow, we could explain such population genetic structure with frequent extinction and re-establishment of populations (McCauley & al., 2003; Westberg & Kadereit, 2009). While we lack data about the over-sea dispersal ability of *Anthemis*, we have observed ephemeral populations both in Favignana (Egadi Islands) and in Lampedusa. A vicariant scenario is however plausible, with recent range reduction and/or fragmentation followed by genetic drift being the predominant force that has influenced the genetic variation of the populations studied. During the last glacial maximum (LGM; c. 19-22 ka; Yokoyama & al., 2000) the Egadi islands (up to Marettimo; Agnesi & al., 1993) were connected to Sicily and the overall coastline was about 120-130 m lower than today (Yokoyama & al., 2000; Thiede, 1978), so that the remaining islands (Pantelleria, Ustica) were separated from mainland only by shallow straits. Only Lampedusa, which was connected to the African continent, was quite isolated from Sicilian mainland. We may therefore suppose high gene flow among at least Egadi, Sicily, Ustica and Pantelleria populations of *A. secundiramea*. The rising in sea level that followed the end of the last glaciation must have caused the extinction of some populations and the reduction on size of others, which have remained isolated from each other. Different outcomes of genetic drift in each of the remaining isolates have finally determined the observed pattern. A similar example in the Mediterranean region is provided by the Aegean *Nigella arvensis* alliance (Bittkau & Comes, 2005, 2009), whose lineages are supposed to have evolved *in situ* as a result of multiple fragmentation events after the last glaciation.

The lower genetic diversity in the two volcanic islands Ustica and Pantelleria may be a still visible signal of a founder effect, which occurred during the LGM or even earlier, but not before the Late Pleistocene, as the two islands appeared (de Vita & al., 1998; Wallmann & al., 1988). The high number of singletons in Lampedusa, on the other hand, can be related to an older colonization of the island and a prolonged isolation, which allowed the accumulation of genetic variation over generations (Stuessy & al., 2006). Moreover, as the populations in the island of Lampedusa are at the geographical margin of the range of *A. secundiramea*, effects such as 'surfing' (i.e. spread of rare alleles determined by rapid spatial expansion) could have promoted a relative accumulation of genetic variation (Excoffier & Ray, 2008).

The N African taxa in a mainland system

In contrast to the pattern observed for *A. secundiramea*, the N African taxa show a significant pattern of isolation-by-distance, and measures of genetic differentiation, both at cpDNA (with N_{ST} significantly different from G_{ST}; Table 4.2) and AFLP level (Φ_{PT} of 0.054), revealing a geographical structure in the genetic variation. Such patterns are observed when populations systems are old enough to have reached the migration-drift equilibrium (see Kadereit & al., 2005) and our assumed age for the whole group of N African taxa (c. 5 Ma; Lo Presti & Oberprieler, 2009; see also chapter 2 and chapter 3) agrees with this idea. However, when looking at the spatial distribution of the chloroplast haplotypes, the different species appear to be indistinguishable by sharing the same haplotypes (Fig. 4.1). Moreover, the BI tree (Fig. 4.3) shows that the haplotypes found in the three taxa here analysed are not separated from those belonging to other N African *Anthemis* species. There is of course the possibility that the species boundaries among the N African taxa here analysed have to be considered not existent. However, even if each species is characterised by a considerable amount of morphological variation (Oberprieler, 1998), they can be morphologically distinguished from each other, at least in the peripheral margins of their range, on the basis of the characters described in the materials & methods section (see above). Thus, given that we are really dealing with morphologically distinct species, the question arises how such a scenario can be interpreted. Sharing of haplotypes across species boundaries was observed in many other plant groups, such as the Mediterranean *Senecio* complex (Comes & Abbott, 2001) or *Coreopsis* (Mason-Gamer & al., 1995) and has been explained by invoking incomplete lineage sorting or reticulation. It is notoriously difficult to disentangle these two factors, considering also that the observed current situation may result from the combination of both. In the first case, the maintenance of ancestral haplotypes through successive speciation events is required (Comes & Abbott, 2001; Funk & Omland, 2003), and may be caused by a fast radiation. As in the case of the Mediterranean *Senecio* complex (Comes & Abbott, 2001), this process could be related to the adaptation to a diversity of new ecological situations, a process known as 'adaptive radiation'. In the case of the N African taxa of *Anthemis*, we can detect a very low overlap among the modelled distribution of the three taxa according to climatic variables (Table 4.3), thus showing that they have different climatic requirements.

Reticulate evolution, on the other hand, can be assumed when distinct species come in contact and hybridize. It is well documented that N Africa was regularly affected by alternating humid and hyperarid phases during the mid-Pliocene to Pleistocene (Street & Gasse, 1981; Quezel & Barbero, 1993). Moreover, in the N African regions, only N Tunisia (Cap Bon) and NE Algeria (Medail & Diadema, 2009) may have acted as refugia during the glaciations of Pleistocene. It is therefore plausible to hypothesize that previously distinct species, which have had enough time to harbour many different haplotypes, have migrated in concomitance with climatic oscillations and have therefore got in secondary contact with each other. Alleles from each species may have penetrated the gene pool of the others, leading to the actual observed scenario.

Acknowledgments

This research was supported by the Bavarian Research Foundation (BFS), by the German Academic Exchange Service (DAAD), by the German Research Foundation (grant OB 155/7-1) and by the SYNTHESYS project of the EU to RMLP and CO (AT-TAF-1618, AT-TAF-1731). We would like to thank P. Hummel for his technical support in the laboratory at the University of Regensburg, everybody who joined the authors in the sampling excursions, and E. Véla (Montpellier) and A. Saatkamp (Marseille) for providing silica-gel dried material from Algeria and France.

Chapter 5

General Discussion

Biogeography and phylogeography: an introduction

Biogeography, as the 'science that attempts to describe and interpret the geographic distribution of organisms' (Avise, 2004: p. 893), has for a long time suffered from the chasm between its two main disciplines, i.e. historical and ecological biogeography. While historical biogeographers have tried to interpret the distribution of organisms as a consequence of evolutionary and geological processes in the past, ecological biogeographers have concentrated on physical and biological factors that could explain 'what lives where, and why' (Parenti & Humphries, 2004: p. 899). However, since 'earth and life evolve together' (Croizat, 1964: p. 858), an integration of both disciplines is needed and desirable (Avise, 2004, Lomolino & Heaney, 2004; Parenti & Humphries, 2004; Wiens & Donoghue, 2004). An integrative historical biogeography, therefore, should involve the geological, geographical, and ecological aspects and should 'quantify the relative importance of these factors in determining the large-scale distribution of clades' (Wiens & Donoghue, 2004: p. 642).

Phylogeography, formally introduced about 20 years ago (Avise & al., 1987), is a field of study concerned with 'the principles and processes governing the geographical distribution of genealogical lineages, especially those at the intraspecific level' (Avise, 1998: p. 371). As it focuses on biogeography within species or among closely related species, it is conceptually positioned between macro- and microevolution. By having successfully expanded population genetic theory and questions into the biogeographical realm, it has provided the basis for a promising bridge between ecological and historical aspects of biogeography (Riddle & Hafner, 2004).

While the integration of both ecological and historical biogeography sheds light to the evolutionary history of taxa (macroevolution, top-down approach), the field of phylogeography investigates the first steps of speciation processes and populations differentiation (microevolution, bottom-up approach). In this thesis, these two disciplines were used in concert to gain further understanding of the evolution of biodiversity in the Mediterranean Basin.

A macroevolutionary approach: searching for an explanation of the diversification in the mediterranean-type regions

The results shown in Chapter 2 and 3 have evidenced that the closely related genera *Anthemis* and *Cota* have radiated in the Mediterranean area in the last 10 Ma. While geological processes seem to have contributed to this radiation in a quite constant manner, the major climatic events of the last 10 Ma have played an important role in the evolutionary history of these taxa. Between 12 and 7 Ma a trend in increasing aridification (Ivanov & al., 2002; Fortelius & al., 2006; Van Dam, 2006) has probably triggered a first radiation of the group in the circum-Mediterranean area. Later, the stabilization of summer drought at about 3 Ma (Thompson, 2005) and the onset of glacial-interglacial oscillations at c. 2.4 Ma (Bertoldi & al., 1989) may have resulted in a new increase in the number of climatic vicariance events (see Fig. 3.6).

In order to evaluate whether an evolutionary radiation has occurred, which in turn can be adaptive or not, a phylogenetic framework with extensive species representation is needed. The almost complete phylogeny of *Anthemis*, with 75% of the total species sequenced, has shed some light to the factors that promoted speciation, and to whether these factors have an adaptive value or not (Savolainen & Forest, 2005). A radiation requires a high species production and can be defined as 'adaptive' only if the diversification is related to a particular adaptation to some environmental features. Cichlid fishes or Darwin´s finches are classical examples of adaptive radiations.

Moreover, when a radiation is rapid, it has been related to the lack of high support in the reconstruction of phylogenies (Hopper & Gioia, 2004). Examples are *Bomarea* in South America (Alzate & al., 2008), *Eucalyptus* (Steane & al., 2002) or *Acacia* (Miller & Bayer, 2001) in Australia: despite the use of sequence data from many nuclear and chloroplast DNA regions, the relationships recovered in these genera remained weakly supported. The observed high diversification rate in *Anthemis* in the last 10 Ma, which has been primarily related to climatic events, coupled with the lack of well-supported basal nodes both in ITS and cpDNA datasets, suggests that *Anthemis* represents an example of both rapid and adaptive radiation.

An adaptive radiation over the last 10 Ma was already observed for other plant groups widespread in the Mediterranean area: *Ranunculus*, for example, occurs with various life forms from the lowlands to the highest mountains (Paun & al., 2005) and its diversification into main ecological clades is reconstructed to have taken place in the late Miocene. It is however quite interesting to note that explosive recent, adaptive radiations have been a major evolutionary event in all mediterranean-climate hotspots of the world.

Verboom and colleagues (2009), for example, have found out that in the succulent karoo, one of the two South African biomes of the CFR, the flora is not older than 17 Ma, with the majority of lineages being younger than 10 Ma. Its appearance has been associated with the increasing aridification of this region, which started in Late Miocene (Verboom & al., 2009). One of the most spectacular examples of adaptive radiation in this region is the family Aizoaceae, which is suggested to have given rise to more than 1,500 species in the last 3.8-8.7 Ma (Klak & al., 2004).

In south Western Australia the increase in speciation rate in the Late Miocene/ Early Pliocene has been related to the climatic fluctuations as Australia became more arid (Hopper & al., 2009; Hopper & Gioia, 2004). In California, Calsbeek & al. (2009) stressed the importance of recent geological and climatological events, i.e., the orogeny of the Sierra Nevada, Coast and Transverse ranges between 5 and 2 Ma, as well as the beginning of aridification about 2.8 Ma, to species diversification and the overall biodiversity of the area. In Chile, the diversification of herbaceous lineages has been interpreted as response to the aridification, which began in the Late Tertiary (Cowling & al., 1996a).

It emerges that the rapid, recent radiations occurred in the mediterranean-type regions are related to a trend in increasing aridification, which culminates with the onset of the typical Mediterranean climate, and which started not earlier than the Late Miocene. Aridity, considered already by Stebbins (1950, 1952) as stimulus for evolution, has been one of the main factors for adaptive radiations to occur, and the genus *Anthemis* in the Mediterranean Basin represents one of the numerous examples of this kind of radiation.

Aridity alone, however, can not explain the high number of taxa in the five Mediterranean regions, which are even hot-spots of biodiversity in the world (Myers & al., 2000). Moreover, it can not explain why CFR and south Western Australia host the highest biodiversity among the five regions (Cowling & al., 1996a). The events underlying the genesis of this diversity have been the subject of considerable discussion in the literature. They range from geomorphic evolution (Cowling & al., 2009) and adaptation to different physical environments (Van der Niet & Johnson, 2009), to pollinator specialization (Waterman & al., 2009), to post-fire regeneration modes (Van der Niet & Johnson, 2009). Nevertheless, as all these factors that drive speciation appear to be not unique to these regions, there is increasing awareness that the origin of species richness can not just be clarified with simplistic explanations.

Linder (2008) argued that it is possible to make a distinction between recent, rapid radiations and older, mature radiations, and that both can result in substantial species diversity. He considered that mature radiations are found in areas of tectonic and climatic stability and are the signature of the absence of dramatic extinction events during the Neogene. A comparison of the ages of the two south African biomes (fynbos and succulent karoo) lineages across 17 groups of plants (Verboom & al., 2009) has shown that the exceptional high diversity of CFR might be the result of a combination of mature radiations (in the fynbos) and recent and rapid radiations (in the succulent karoo). The fynbos-endemic family Bruniaceae, for example, was inferred to be 59.7-99.5 Ma old (Quint & Classen-Bockhoff, 2004). The rich Australian flora, on the other side, can be described as mature radiations (the family Haemodoraceae, for example, is reconstructed to be 90-81 Ma old; Hopper & al., 2009). Low extinction rates related to the relative climatic and geomorphic stability during the Quaternary have allowed the survival of ancient taxa, with the result that actual floras are rich both in old and young lineages (Cowling & al., 2009; Verboom & al., 2009). Lineage diversity has been accumulating for a long time, and has resulted in the current high species diversity.

Recent radiations, on the other side, are associated with new habitats, which are mostly the result of geotectonic changes (Linder, 2008). In California and in the Mediterranean basin relatively extreme glacial conditions in the Quaternary are responsible of higher extinction rates of the Tertiary flora (Cowling & al., 1996a), with the result that their current flora is resulting primarily from recent radiations. These, in turn, are considered to be related to new space becoming available, and *Anthemis*, as a typical member of the Mediterranean area vegetation, represents a classic example for both a recent and adaptive radiation. California and the Mediterranean basin are characterized by the highest habitat heterogeneity among all mediterranean-climate areas (Cowling & al., 2009), which is very recent and still evolving rapidly. Tectonic forces induced mountain building as recently as the Pliocene and Quaternary geomorphic processes, also related to the sea-level fluctuations associated to the glaciations cycles, have shaped contemporary landscapes. Habitat heterogeneity has been, and still is, a factor that enhances differentiation and thus radiation (Rosenzweig, 1995).

A microevolutionary approach: a matter of scale

As the Mediterranean basin hosts a remarkable high topographic heterogeneity, the question arises why geographical processes seem to have had a relative lower influence than climatological ones in the evolutionary history of the genus *Anthemis* (see Fig. 3.6).

As already discussed in chapter 3, the relatively large predefined areas adopted in the DIVA analysis (see Fig. 3.1) may hide allopatric patterns on a finer geographical scale, which in turn may result in a reduction of the effective number of geographical vicariance events. Chapter 4, therefore, aimed at a clarification of this issue, by adopting a phylogeographic approach and focussing on a more local scale, namely the *A. secundiramea* group distributed across the Sicilian Channel (N Africa, Sicily and its surrounding islands and islets).

When dealing with the phylogeny of the whole genus *Anthemis*, thus considering the circum-Mediterranean area (from the Mediterranean Basin to Near East Asia, Iraq and Iran), only one geographical vicariance event can be recognized within the *Anthemis secundiramea* group, namely between the areas including Sicily and the Maltese archipelago on one side, N Africa on the other (see Fig. 3.4). At a much smaller spatial scale, which includes only the regions across the Sicilian Channel in the Central Mediterranean, new, numerous allopatric distribution patterns are revealed, which were interpreted as sympatric distributions at a larger scale. Within Sicily, *A. secundiramea* is widespread along the western and northern coasts, while *A. muricata* is localized to the arid hills of the inland. Moreover, *A. secundiramea* is not found in the Maltese archipelago, where the endemic *A. urvilleana* is distributed. Within N Africa, the three taxa investigated are allopatrically distributed: *A. ubensis* inhabits the mountains of N Tunisia, *A. confusa* is centred on the semi-desert areas of C and S Tunisia, and *A. glareosa* is restricted to the plains along the coasts of N Libya and S Tunisia.

It emerges that the habitat heterogeneity can be recognized more at a local than at a global scale. It seems therefore necessary to distinguish between geological events, which occur on a global scale, and geographical events, which require a more local scale to be conceived. While the geological events of the last 10 Ma have had a constant influence on the radiation of *Anthemis* in the Mediterranean area, finer-scale geographical events seem to be mainly related to the sea-fluctuation levels correlated to the glaciations of the Quaternary (see chapter 4).

Furthermore, chapter 4 showed how the observed allopatric distributions can be related both to vicariance and dispersal. The distribution patterns of *A. secundiramea* in Sicily and of *A. urvilleana* in the Maltese archipelago well illustrate a speciation event linked to geographical vicariance, which occurred as the two land masses became definitively isolated from each other in the present interglacial. The N African taxa, on the other side, are a good example of speciation events related to range expansion/contraction, which in turn may have been driven by ecological adaptations to new conditions.

A phylogeographic approach, therefore, considers other mechanisms than just vicariance that influence the evolutionary and geographical history of genetic lineages. Aiming to incorporate a more inclusive perspective than historical biogeography, phylogeography seeks to explore both vicariance and dispersal in the history of lineages, and events such as range expansion or contraction, which do not have any analytical standing in historical biogeography, are legitimate and important components of phylogeography (Riddle & Hafner, 2004).

Conclusion: linking macro- and microevolution

In the persistent debate whether macroevolutionary trends are governed by the principles of microevolution, we agree that macroevolution is not simply 'the product of microevolution writ large' (Carroll, 2001: 669), especially in the case of rapid speciation, which is considered to produce a discontinuity between macro- and microevolution (Erwin, 2000). As we have seen above and in chapter 3, understanding macroevolution requires the integration of ecology and the role of history in shaping the diversification or decline of lineages (Reznick & Ricklefs, 2009). However, as discussed above and in chapter 4, a phylogeographical approach, which sheds light on local mechanisms of speciation, is needed to better understand the first steps of the origin of new species, and constitutes therefore an essential piece in the jigsaw puzzle of the evolution of biodiversity.

Thus, scale-boundaries between evolutionary disciplines seem to be inadequate (Carroll, 2001), as both macro- and microevolutionary approaches should be involved to understand patterns and processes in the evolution of biodiversity. Past and present are inexorably linked in any study of evolution (Thompson, 2005), and are both requested to gain the ability to predict the impacts of future environmental changes on biodiversity. Improving our understanding of the genetic, ecological and historical causes of the origin of biodiversity will finally facilitate conservation efforts in biodiversity hotspots worldwide.

Summary

The five mediterranean-climate regions of the world occupy less than 5% of the Earth´s surface but harbour c. 20% of the world´s total plant species. Among them, the Mediterranean Basin is the biggest in terms of total area covered and includes 11 hotspots defined on the basis of plant endemism and richness. This region seems particularly suitable as a model system in which to integrate the study of species divergence (macroevolution) with that of population differentiation (microevolution), as it has been evidenced that both ecological specialization and geographical isolation have been primary determining factors to explain such a high biodiversity.

In the Mediterranean area, the genus *Anthemis* L. (Asteraceae, Anthemideae) provides a suitable plant group with which to link both the macro- and the microevolutionary approaches. As it was reconstructed to have diverged from its close relatives between 10 to 15 Ma, it acts as a suitable proxy for the reconstruction of the biogeographical and climatological history of the Mediterranean area, spanning the transition from the subtropical climate of the Early Miocene to the typical Mediterranean environment of the present. On the other side, it includes many closely related groups of species, such as the *Anthemis secundiramea* group widespread across the Sicilian Channel, which provide suitable models to study the role of geographical and/or ecological diversification on a more local scale.

In the first part (Chapter 2) the monophyly of the two closely related genera *Anthemis* and *Cota* and the relationships among the infrageneric groups were examined, on the basis of a comprehensive sampling including c. 75% of all known species belonging to the two genera. A molecular phylogeny was based on sequence information from two plastid regions (*psbA-trnH* and *trnC-petN* spacer regions of the chloroplast DNA) and one nuclear marker [internal transcribed spacer (ITS) of the nuclear ribosomal DNA (nrDNA)] and was supported by 25 micro-morphological features. A relaxed Bayesian clock was adopted to estimate the age of the diversification of *Anthemis* and *Cota*, and the calibration used a combination of secondary age estimates, fossil pollen and geological events. The results showed that *Anthemis* s.l. is not monophyletic and that the traditional infrageneric classification is not in congruence with the phylogeny of the genus. An evaluation of the temporal diversification of *Anthemis* s.l. based on lineages-through-time (LTT) plots for the entire genus and its subdivisions showed that the rate of diversification varies among clades due to their predominant life-forms (annuals, perennials) and habitats.

The goals of the second part (Chapter 3) were to contribute to an assessment of the relative importance of geographical vs. climatological forces involved in the diversification of the genus. The phylogeny of the genus *Anthemis* was obtained from a maximum-likelihood-analysis based on nrDNA-ITS sequence data, and the chronology of diversification was derived using a penalized likelihood approach. The reconstruction of the spatial diversification of the genus was based on a dispersal/vicariance (DIVA) analysis. Eco-climatological niche differentiation was inferred by optimizing 19 bioclimatic variables onto the phylogeny. A multi-dimensional hypervolume, proposed as a representation of the eco-climatological niche and defined by the combination of ranges for all bioclimatic variables, was calculated for each taxon and each internal node. To identify "eco-climatological vicariance" events in the phylogeny, the pairwise overlap among hypervolumes of sister groups was calculated. Finally, the temporal and clade-wise relative importance of geographical vs. eco-climatological vicariance events was estimated. The results have evidenced that the closely related genera *Anthemis* and *Cota* have radiated in the Mediterranean area in the last 10 Ma. While geological processes seem to have contributed to this radiation in a quite constant manner, climatic forces seem to have played an important role in two phases of the radiation process: at around 9 Ma when the area experienced the onset of a trend towards aridification, and during the last 3.5 Ma, with the establishment of the typical Mediterranean climate and the influence of Pleistocene climate oscillations.

The third part (Chapter 4) aimed to the identification of the role of geographical vs. climatological differentiation processes at a local scale, by focussing on the *A. secundiramea* group, made up of six closely related species distributed across the Sicilian Channel (N Africa, Sicily and its surrounding islands and islets). The geographical distribution of the genetic variability and differentiation was studied by using both chloroplast DNA variation (sequences of the two spacer regions *psbA-trnH* and *trnC-petN*) and AFLP fingerprinting. The potential distributions of the six species according to climatic factors were modelled using Maxent. The results identified three well-defined groups of taxa: the first included all populations located in N Africa, which showed a significant pattern of isolation-by-distance. Their sharing of the same haplotypes has been related to range expansion/contraction, which in turn might be linked to the climatic oscillations of the Quaternary. The second group comprised all populations belonging to *Anthemis secundiramea*: the absence of genetic structure suggested an actual or at least recent gene flow among them. The third group included the populations belonging to *A. secundiramea* subsp. *urvilleana*, endemic to Malta archipelago. While the hybridization with *A. arvensis* or *A. peregrina* was hypothesised for four populations, the sharing of a haplotype between one population and *A. muricata*, endemic to Sicily, has been interpreted as a remnant of the ancient connection between the Maltese archipelago and south-Eastern Sicily.

This thesis has shown that both macro- and microevolutionary approaches should be involved to understand patterns and processes in the evolution of biodiversity. Understanding macroevolution requires the integration of ecology and the role of history in shaping the diversification or decline of lineages (chapter 2 and 3). A phylogeographical approach (chapter 4), which sheds light on local mechanisms of speciation, is needed to better understand the first steps of the origin of new species, and constitutes therefore an essential piece in the jigsaw puzzle of the evolution of biodiversity.

References

Agnesi, V., Macaluso, T., Orrù, P. & Ulzega, P. 1993. Paleogeografia dell'arcipelago delle Egadi (Sicilia) nel Pleistocene sup.-Olocene. *Naturalista Siciliano* S. IV, XVII: 3-22.

Agnew, A.D.Q. & Agnew, A. 1994. *Upland Kenya wild flowers*. East Africa Natural History Society, Nairobi.

Agusti, J., Sanz de Siria, A. & Garces, M. 2003. Explaining the end of the hominoid experiment in Europe. *Journal of Human Evolution* 45: 145-153.

Albach, D.C., Martínez-Ortega, M.M. & Chase, M.W. 2004. *Veronica*: parallel morphological evolution and phylogeography in the Mediterranean. *Plant Systematics and Evolution* 246: 177-194.

Alexander, D. 1988. A review of the physical geography of Malta and its significance for tectonic geomorphology. *Quaternary Science Reviews* 7: 41-53.

Alzate, F., Mort, M.E. & Ramirez, M. 2008. Phylogenetic analyses of *Bomarea* (Alstroemeriaceae) based on combined analyses of nrDNA ITS, *psbA-trnH*, *rpoB-trnC* and *matK* sequences. *Taxon* 57: 853-862.

Avise, J.C. 1998. The history and purview of phylogeography: a personal reflection. *Molecular Ecology* 7: 371-379.

Avise, J.C. 2000. *Phylogeography – The history and formation of species*. Cambridge, London.

Avise, J.C. 2004. What is the field of biogeography, and where is it going? *Taxon* 53: 893-898.

Avise, J.C., Arnold, J., Ball, R.M., Bermingham, E., Lamb, T., Neigel, J.E., Reeb, C.A. & Saunders, N.C. 1987. Intraspecific phylogeography: the mitochondrial DNA bridge between population genetics and systematics. *Annual Review of Ecology and Systematics* 18: 489-522.

Axelrod, D.I. 1975. Evolution and biogeography of the Madrean-Tethyan sclerophyllous vegetation. *Annals of the Missouri Botanical Garden* 62: 280-334.

Barraclough, T.G. & Reeves, G. 2005. The causes of speciation in flowering plant lineages: species-level DNA trees in the African genus *Protea*. Pp. 31-46.

in: Bakker, F.T., Chathrou, L.W., Gravendeel, B. and Pelser, P.B. (ed.), *Plant species-level systematics: new perspectives on pattern & process*. A.R.G. Gantner Verlag, Ruggell, Liechtenstein.

Bena, G., Lejeune, B., Prosperi, J.M. & Olivieri, I. 1998. Molecular phylogenetic approach for studying life-history evolution: the ambiguous example of the genus *Medicago* L. *Proceedings of the Royal Society B: Biological Sciences* 265: 1141-1151.

Benedì i Gonzàlez, C. 1987. *Revisiò biosystemàtica del gènre Anthemis L. a la Penìnsula Ibèrica i les Illes Balears*. Ph. D. Thesis. Barcelona.

Bergh, N.G. & Linder, P.H. 2009. Cape diversification and repeated out-of-southern-Africa dispersal in paper daisies (Asteraceae-Gnaphalieae). *Molecular Phylogenetics and Evolution* 51: 5-18.

Bertoldi, R., Rio, D. & Thunell, R. 1989. Pliocene-Pleistocene vegetational and climatic evolution of the south-central Mediterranean. *Palaeogeography, Palaeoclimatology, Palaeoecology* 72: 263-275.

Bittkau, C. & Comes, H.P. 2005. Evolutionary processes in a continental island system: molecular phylogeography of the Aegean *Nigella arvensis* alliance (Ranunculaceae) inferred from chloroplast DNA. *Molecular Ecology* 14: 4065-4083.

Bittkau, C. & Comes, H.P. 2009. Molecular inference of a Late Pleistocene diversification shift in *Nigella* s. lat. (Ranunculaceae) resulting from increased speciation in the Aegean archipelago. *Journal of Biogeography* 36: 1346-1360.

Blondel, J. 2006. The 'design' of Mediterranean landscapes: a millennial story of humans and ecological systems during the historic period. *Human Ecology* 34: 713-729.

Boissier, P.E. 1849. *Diagnoses plantarum orientalium novarum, Vol. 11*. Paris, France.

Boissier, P.E. 1875. *Anthemis* L. Pp. in: (ed.), *Flora Orientalis, vol. 3*. Genf-Basel.

Bremer, K., Friis, E. & Bremer, B. 2004. Molecular phylogenetic dating of asterid flowering plants shows early Cretaceous diversification. *Systematic Biology* 53: 496-505.

Bremer, K. & Humphries, C.J. 1993. Generic monograph of the Asteraceae-Anthemideae. *Bulletin of the Natural History Museum of London (Botany)* 23: 71–177.

Calsbeek, R., Thompson, J.N. & Richardson, J.E. 2003. Patterns of molecular evolution and diversification in a biodiversity hotspot: the California Floristic Province. *Molecular Ecology* 12: 1021-1029.

Carroll, S.B. 2001. The big picture. *Nature* 409: 669.

Cassar, L.F., Conrad, E. & Schembri, P.J. 2008. The Maltese archipelago. Pp. 297-324. in: Vogiatzakis, I.N., Pungetti, G. and Mannion, A.M. (ed.), *Mediterranean islands landscapes - natural and cultural approaches*. Springer, Netherlands.

Catalano, R., Infuso, S. & Sulli, A. 1995. Tectonic history of the submerged Maghrebian Chain from the Southern Tyrrhenian Sea to the Pelagian Foreland. *Terra Nova* 7: 179-188.

Chester, D.K., Duncan, A.M., Guest, J.E. & Kilburn, C.R.J. 1985. *Mount Etna: the anatomy of a volcano*. Stanford Univ Press. London.

Clemens, S.C., Murray, D.W. & Prell, W.L. 1996. Nonstationary phase of the Plio-Pleistocene asian monsoon. *Science* 274: 943-948.

Clement, M., Posada, D. & Krandall, K.A. 2000. TCS: a computer program to estimate gene genealogies. *Molecular Ecology* 9: 1657-1660.

Coates, D.J. 2000. Defining conservation units in a rich and fragmented flora: implications for the management of genetic resources and evolutionary processes in south-west Australian plants. *Australian Journal of Botany* 48: 329-339.

Collina-Girard, J. 2001. L'Atlantide devant le détroit de Gibraltar? Mythe et géologie. *Earth and Planetary Sciences* 333: 233-240.

Combourieu-Nebout, N. 1993. Vegetation response to upper Pliocene Glacial/Interglacial cyclicity in the Central Mediterranean. *Quaternary Research* 40: 228-236.

Comes, H.P. & Abbott, R.J. 2001. Molecular phylogeography, reticulation, and lineage sorting in Mediterranean *Senecio* sect. *Senecio* (Asteraceae). *Evolution* 55: 1943-1962.

Comes, H.P. & Kadereit, J.W. 2003. Spatial and temporal patterns in the evolution of the flora of the European Alpine System. *Taxon* 52: 451-462.

Conti, E., Soltis, D.E., Hardig, T.M. & Schneider, J. 1999. Phylogenetic relationships of the silver saxifrages (*Saxifraga*, sect. *Ligulatae* Haworth): implications for the evolution of substrate specificity, life histories, and biogeography. *Molecular Phylogenetics and Evolution* 13: 536-555.

Cooper, S.J.B., Hinze, S., Leys, R., Watts, C.H.S. & Humphreys, W.F. 2002. Islands under the desert: molecular systematics and evolutionary origins of stygobitic water beetles (Coleoptera: Dytiscidae) from central western Australia. *Invertebrates Systematics* 16: 589-598.

Cowie, R.H. & Holland, B.S. 2006. Dispersal is fundamental to biogeography and the evolution of biodiversity on oceanic islands. *Journal of Biogeography* 33: 193-198.

Cowling, R.M., MacDonald, I.A.W. & Simmons, M.T. 1996b. The Cape Peninsula, South Africa: physiographical, biological and historical background to an extraordinary hot-spot of biodiversity. *Biodiversity and Conservation* 5: 527-550.

Cowling, R.M., Proches, S. & Partridge, T.C. 2009. Explaining the uniqueness of the Cape flora: incorporating geomorphic evolution as a factor for explaining its diversification. *Molecular Phylogenetics and Evolution* 51: 64-74.

Cowling, R.M., Rundel, P.W., Lamont, B.B., Kalin Arroyo, M. & Arianoutsou, M. 1996a. Plant diversity in mediterranean-climate regions. *Trends in Ecology & Evolution* 11: 362-366.

Crisp, M.D., Arroyo, M.T.K., Cook, L.G., Gandolfo, M.A., Jordan, G.J., McGlone, M.S., Weston, P.H., Westoby, M., Wilf, P. & Linder, H.P. 2009. Phylogenetic biome conservatism on a global scale. *Nature* 458: 754-756.

Croizat, L. 1964. *Space, time, form: The biological synthesis.* Published by the author, Caracas, Venezuela.

Daget, P. 1977a. Le bioclimat mediterraneen: caracteres generaux, modes de caracterisation. *Plant Ecology* 34: 1-20.

Daget, P. 1977b. Le bioclimat Mediterraneen: analyse des formes climatiques par le systeme d'Emberger. *Plant Ecology* 34: 87-103.

Dallman, P.P. 1998. Plant life in the world's Mediterranean climates. *Oxford, Oxford University Press.*

Datson, P., Murray, B. & Steiner, K. 2008. Climate and the evolution of annual/perennial life-histories in *Nemesia* (Scrophulariaceae). *Plant Systematics and Evolution* 270: 39-57.

de Vita, S., Laurenzi, M.A., Orsi, G. & Voltaggio, M. 1998. Application of ^{40}Ar/^{39}Ar and ^{230}Th dating methods to the chronostratigraphy of quaternary basaltic volcanic areas: the Ustica island case history. *Quaternary International* 47-48: 117-127.

deMenocal, P.B. 1995. Plio-Pleistocene african climate. *Science* 270: 53-59.

Demesure, B., Sodzi, N. & Petit, R.J. 1995. A set of universal primers for amplification of polymorphic non-coding regions of mitochondrial and chloroplast DNA in plants. *Molecular Ecology* 4: 129-134.

Doyle, J.J. & Doyle, J.S. 1987. A rapid DNA isolation procedure for small quantatities of fresh leaf tissue. *Phytochemical Bullettin* 19: 11-15.

Drummond, A.J., Ho, S.Y.W., Phillips, M.J. & Rambaut, A. 2006. Relaxed phylogenetics and dating with confidence. *PLoS Biology* 4: e88 doi:10.1371/journal.pbio.0040088.

Drummond, A.J., Ho, S.Y.W., Rawlence, N. & Rambaut, A. 2007. A Rough Guide to BEAST 1.4. Available at: http://beast.bio.ed.ac.uk/.

Drummond, A.J. & Rambaut, A. 2007. BEAST: Bayesian evolutionary analysis by sampling trees. *BMC Evolutionary Biology* 7: 214.

Drummond, C.S. 2008. Diversification of *Lupinus* (Leguminosae) in the western new world: derived evolution of perennial life history and colonization of montane habitats. *Molecular Phylogenetics and Evolution* 48: 408-421.

Duggen, S., Hoernle, K., van den Bogaard, P., Rupke, L. & Morgan, J.P. 2003. Deep roots of the Messinian Salinity Crisis. *Nature* 422: 602–606.

Duncan, A.M., Chester, D.K. & Guest, J.E. 1984. The Quaternary stratigraphy of mount Etna, Sicily: the effects of differing palaeoenvironments on styles of volcanism. *Bulletin of Volcanology* 47: 497-516.

Edwards, C.E., Soltis, D.E. & Soltis, P.S. 2008. Using patterns of genetic structure based on microsatellite loci to test hypotheses of current hybridization, ancient hybridization and incomplete lineage sorting in *Conradina* (Lamiaceae). *Molecular Ecology* 17: 5157-5174.

Edwards, S.V. 2009. Is a new and general theory of molecular systematics emerging? *Evolution* 63: 1-19.

Eig, A. 1938. Taxonomic studies on the oriental species of the genus *Anthemis*. *Palestine Journal of Botany, Jerusalem ser.* 1: 161-224.

Eldenäs, P.K. & Linder, H.P. 2000. Congruence and complementarity of morphological and *trnL-trnF* sequence data and the phylogeny of the african Restionaceae. *Systematic Botany* 25: 692-707.

Elith, J., Graham, C.H., Anderson, R.P., Dudik, M., Ferrier, S., Guisan, A., Hijmans, R.J., Huettmann, F., Leathwick, J.R., Lehmann, A., Li, J., Lohmann, L.G., Loiselle, B.A., Manion, G., Moritz, C., Nakamura, M., Nakazawa, Y., McC. M. Overton, J., Townsend Peterson, A., Phillips, S.J., Richardson, K., Scachetti-Pereira, R., Schapire, R.E., Soberon, J., Williams, S., Wisz, M.S. & Zimmermann, N.E. 2006. Novel methods improve prediction of species' distributions from occurrence data. *Ecography* 29: 129-151.

Ellison, N.W., Liston, A., Steiner, J.J., Williams, W.M. & Taylor, N.L. 2006. Molecular phylogenetics of the clover genus (*Trifolium*-Leguminosae). *Molecular Phylogenetics and Evolution* 39: 688-705.

Erwin, D.H. 2000. Macroevolution is more than repeated rounds of microevolution. *Evolution & Development* 2: 78-84.

Evans, M.E.K., Hearn, D.J., Hahn, W.J., Spangle, J.M. & Venable, D.L. 2005. Climate and life-history evolution in evening primroses (*Oenothera*, Onagraceae): a phylogenetic comparative analysis. *Evolution* 59: 1914-1927.

Excoffier, L., Laval, G. & Schneider, S. 2005. Arlequin ver. 3.0: an integrated software package for population genetics data analysis. *Evolutionary Bioinformatics Online* 1: 47-50.

Excoffier, L. & Ray, N. 2008. Surfing during population expansions promotes genetic revolutions and structuration. *Trends in Ecology & Evolution* 23: 347-351.

Fauquette, S., Suc, J.-P., Guiot, J., Diniz, F., Feddi, N., Zheng, Z., Bessais, E. & Drivaliari, A. 1999. Climate and biomes in the West Mediterranean area during the Pliocene. *Palaeogeography, Palaeoclimatology, Palaeoecology* 152: 15-36.

Fauquette, S., Suc, J.P., Bertini, A., Popescu, S.M., Warny, S., Bachiri Taoufiq, N., Perez Villa, M.J., Chikhi, H., Feddi, N. & Subally, D. 2006. How much did climate force the Messinian salinity crisis? Quantified climatic conditions from pollen records in the Mediterranean region. *Palaeogeography, Palaeoclimatology, Palaeoecology* 238: 281-301.

Fedorov, A. 1961. *Anthemis* L. Pp. 8-72. in: Komarov, V.L. (ed.), *Flora of the U.S.S.R.: Volume XXVI – Compositae.* Akad. Nauk. SSSR, Moskva-Leningrad.

Felsenstein, J. 1985. Confidence limits of phylogenies: an approch using the bootstrap. *Evolution* 39: 783-791.

Fernandes, R.B. 1975. Identification, typification, affinités et distribution géographique de quelques taxa européens du genre *Anthemis* L. *Anales del Instituto Botánico A. J. Cavanilles* 32: 1409-1488.

Fernandes, R.B. 1976. *Anthemis* L. Pp. 145-159. in: Tutin, T.G., Heywood, V.H., Burges, N.A., Moore, D.M., Valentine, D.H., Walters, S.M., Webb, D.A. (ed.), *Flora Europaea: Volume 4.* Cambridge.

Fiz, O., Valcárcel, V. & Vargas, P. 2002. Phylogenetic position of Mediterranean Astereae and character evolution of daisies (*Bellis*, Asteraceae) inferred from nrDNA ITS sequences. *Molecular Phylogenetics and Evolution* 25: 157-171.

Ford, K.A., Ward, J.M., Smissen, R.D., Wagstaff, S.J. & Breitwieser, I. 2007. Phylogeny and biogeography of *Craspedia* (Asteraceae: Gnaphalieae) based on ITS, ETS and *psbA-trnH* sequence data. *Taxon* 56: 783-794.

Fortelius, M., Eronen, J., Liu, L., Pushkina, D., Tesakov, A., Vislobokova, I. & Zhang, Z. 2006. Late Miocene and Pliocene large land mammals and climatic changes in Eurasia. *Palaeogeography, Palaeoclimatology, Palaeoecology* 238: 219-227.

Francisco-Ortega, J., Barber, J.C., Santos-Guerra, A., Febles-Hernandez, R. & Jansen, R.K. 2001. Origin and evolution of the endemic genera of Gonosperminae (Asteraceae: Anthemideae) from the Canary Islands: evidence from nucleotide sequences of the internal transcribed spacers of the nuclear ribosomal DNA. *American Journal of Botany* 88: 161-169.

Francisco-Ortega, J., Santos-Guerra, A., Hines, A. & Jansen, R.K. 1997. Molecular evidence for a Mediterranean origin of the Macaronesian endemic genus *Argyranthemum* (Asteraceae). *American Journal of Botany* 84: 1595-1613.

Frankham, R. 1997. Do island populations have less genetic variation than mainland populations? *Heredity* 78: 311-327.

Frazzetta, G. & Villari, L. 1981. The feeding of the eruptive activity of Etna volcano. The regional stress field as a constraint to magma uprising and eruption. *Bulletin of Volcanology* 44: 269-282.

Fuertes Aguilar, J., Rossello, J.A. & Nieto Feliner, G. 1999. Nuclear ribosomal DNA (nrDNA) concerted evolution in natural and artificial hybrids of *Armeria* (Plumbaginaceae). *Molecular Ecology* 8: 1341-1346.

Funk, D.J. & Omland, K.E. 2003. Species-level paraphyly and polyphyly: frequency, causes, and consequences, with insights from animal mitochondrial DNA. *Annual review of ecology, evolution and systematics* 34: 397-423.

Funk, V.A., Anderberg, A.A., Baldwin, B.G., Bayer, R.J., Bonifacino, M., Breitwieser, I., Brouillet, L., Carbajal, R., Chan, R., Coutinho & al. 2009. Compositae Meta-supertree: The next generation. Pp. 747-777. in: Funk, V.A., Susanna, A., Stuessy, T. and Bayer, R.J. (ed.), *Systematics and Evolution of the Compositae.* IAPT, Vienna, Austria.

Gardiner, W., Grasso, M. & Sedgeley, D. 1995. Plio-pleistocene fault movement as evidence for mega-block kinematics within the Hyblean-Malta Plateau, Central Mediterranean. *Journal of Geodynamics* 19: 35-51.

Gautier, F., Caluzon, G., Suk, J.P. & Violanti, D. 1994. Age et durée de la crise de salinité Messinienne. *Comptes Rendus de l'Académie des Sciences de Paris* 318: 1103–1109.

Geôrgiou, O. 1990. *Biosustêmatikê meletê tês Anthemis tomentosa (Asteraceae) stên Ellada.* Ph. D. Thesis, Univ. Patras.

Ghafoor, A. & Al-Turki, T.A. 1997. A synopsis of the genus *Anthemis* L. (Compositae-Anthemideae) in Saudi Arabia. *Candollea* 52: 457–474.

Giraudi, C. 2004. The Upper Pleistocene to Holocene sediments on the Mediterranean island of Lampedusa (Italy). *Journal of Quaternary Science* 19: 537-545.

Goldblatt, P. & Manning, J.C. 2002. Plant diversity of the Cape region of southern Africa. *Annals of the Missouri Botanical Garden* 89: 281-302.

Goloboff, P.A. 1999. Analyzing large data sets in reasonable times: solutions for composite optima. *Cladistics* 15: 415-428.

Goloboff, P.A., Farris, J.S. & Nixon, K.C. 2008. TNT, a free program for phylogenetic analysis. *Cladistics* 24: 774-786.

Graham, A. 1996. A contribution to the geological history of the Compositae. Pp. 123– 140. in: D. Hind and Beentje, H. (ed.), *Proceedings of the Kew International Compositae Conference 1994, vol. 1.* Royal Botanic Gardens, Kew.

Grant, V. 1971. *Plant speciation.* Columbia University Press. New York and London.

Greger, H. 1969. Flavonoide und Systematik der Anthemideae (Asteraceae). *Naturwissenschaften* 56: 467-468.

Greger, H. 1970. *Flavonoide und Systematik der Anthemideae (Asteraceae).* Ph.D. Thesis, Univ. Graz.

Greger, H. 1977. Anthemideae – chemical review. Pp. 889-941. in: Heywood, V.H., Harborne, J.B. and Turner, B.L. (ed.), *The biology and chemistry of the Compositae.* London.

Greuter, W. 1968. Contributio floristica austro-aegaea 13. *Candollea* 23: 145-150.

Greuter, W., Oberprieler, C. & Vogt, R. 2003. The Euro+Med treatment of *Anthemideae* (Compositae) – generic concepts and required new names. *Willdenowia* 33: 37-43.

Grierson, A.J.C. & Yavin, Z. 1975. *Anthemis* L. Pp. 174-221. in: Davis, P.H. (ed.), *Flora of Turkey and the East Aegean Islands, vol. 5.* Edinburgh.

Guillou, H., Carracedo, J.C., Paris, R. & Torrado, F.J.P. 2004. Implications for the early shield-stage evolution of Tenerife from K/Ar ages and magnetic stratigraphy. *Earth and Planetary Science Letters* 22: 599-614.

Guo, Y.P., Ehrendorfer, F. & Samuel, R. 2004. Phylogeny and systematics of *Achillea* (Asteraceae-Anthemideae) inferred from nrITS and plastid *trnL-F* DNA sequences. *Taxon* 53: 657-672.

Gussone, G. 1845. *Florae Siculae Synopsis.* 2. Napoli.

Hall, T.A. 1999. BioEdit: a user-friendly biological sequence alignment editor and analysis program for Windows 95/98/NT. *Nucleic Acids Symposium Series* 41: 95-98.

Hamilton, M.B. 1999. Four primer pairs for the amplification of chloroplast intergenic regions with intraspecific variation. *Molecular Ecology* 8: 347-525.

Hardy, C.R. & Linder, H.P. 2005. Intraspecific variability and timing in ancestral ecology reconstruction: a test case from the Cape Flora *Systematic Biology* 54: 299-316.

Harling, G. 1950. Embryological studies in the Compositae. I. Anthemideae-Anthemidinae. *Acta Horti Bergiani* 15: 135-168.

Harling, G. 1951. Embryological studies in the Compositae. II. Anthemideae-Chrysantheminae. *Acta Horti Bergiani* 16: 1-56.

Harling, G. 1960. Further embryological and taxonomical studies in *Anthemis* L. and some related genera. *Svensk Botanisk Tidskrift* 54: 572-590.

Hedberg, O. 1957. Afroalpine vascular plants. *Symbolae botanicae Uppsaliensis* 15: 1-411.

Hellwig, F.H. 2004. Centaureinae (Asteraceae) in the Mediterranean. History of ecogeographical radiation. *Plant Systematics and Evolution* 246: 137-162.

Hendy, M.D. & Penny, D. 1989. A framework for the quantitative study of evolutionary trees. *Systematic Zoology* 38: 297-309.

Hershkovitz, M.A., Arroyo, M.T.K., Bell, C. & Hinojosa, L.F. 2006. Phylogeny of *Chaetanthera* (Asteraceae: Mutisieae) reveals both ancient and recent origins of the high elevation lineages. *Molecular Phylogenetics and Evolution* 41: 594-605.

Himmelreich, S., Källersjo, M., Eldenäs, P. & Oberprieler, C. 2008. Phylogeny of southern hemisphere Compositae-Anthemideae based on nrDNA ITS and cpDNA ndhF sequence information. *Plant Systematics and Evolution* 272: 131-153.

Holland, B.R., Clarke, A.C. & Meudt, H.M. 2008. Optimizing automated AFLP scoring parameters to improve phylogenetic resolution. *Systematic Biology* 57: 347 - 366.

Hopper, S.D. & Gioia, P. 2004. The Southwest Australian Floristic Region: evolution and conservation of a global hot spot of biodiversity. *Annual Review of Ecology, Evolution, and Systematics* 35: 623-650.

Hopper, S.D., Smith, R.J., Fay, M.F., Manning, J.C. & Chase, M.W. 2009. Molecular phylogenetics of Haemodoraceae in the Greater Cape and Southwest Australian Floristic regions. *Molecular Phylogenetics and Evolution* 51: 19-30.

Hsü, K.J. 1972. When the Mediterranean dried up. *Scientific American* 227: 27-36.

Hughes, C. & Eastwood, R. 2006. Island radiation on a continental scale: exceptional rates of plant diversification after uplift of the Andes. *Proceedings of the National Academy of Sciences* 103: 10334–10339.

Iranshahr, M. 1982a. Fünf neue Arten der Gattung *Anthemis* (Compositae) aus Iraq und Afghanistan. *Plant Systematics & Evolution* 139: 159-162.

Iranshahr, M. 1982b. Acht neue Arten der Gattung *Anthemis* (Compositae) aus Persien. *Plant Systematics & Evolution* 139: 313-317.

Iranshahr, M. 1986. *Anthemis* L. Pp. 5-44. in: Rechinger, K.H. (ed.), *Flora Iranica: Vol. 158 Compositae VI - Anthemideae*. Akademische Druck- u. Verlagsanstalt, Graz-Austria.

Ivanov, D., Ashraf, A.R., Mosbrugger, V. & Palamarev, E. 2002. Palynological evidence for Miocene climate change in the Forecarpathian Basin (Central Paratethys, NW Bulgaria). *Palaeogeography, Palaeoclimatology, Palaeoecology* 178: 19-37.

Jolivet, L., Augier, R., Robin, C., Suc, J.P. & Rouchy, J.M. 2006. Lithospheric-scale geodynamic context of the Messinian salinity crisis. *Sedimentary Geology* 188-189: 9-33.

Junikka, L., Uotila, P. & Lahti, T. 2006. A phytogeographical comparison of the major Mediterranean islands on the basis of Atlas Florae Europaeae. *Willdenowia* 36: 379-388.

Kadereit, J.W. & Comes, H.P. 2005. The temporal course of alpine plant diversification in the Quaternary. Pp. 117-130. in: F.T. Bakker, L.W. Chathrou, B. Gravendeel and Pelser, P.B. (ed.), *Plant species-level systematics: new perspectives on pattern & process*. A.R.G. Gantner Verlag, Ruggell, Liechtenstein.

Kay, M., Reeves, P., Olmstead, R. & Schemske, D. 2005. Rapid speciation and the evolution of hummingbird pollination in neotropical *Costus* subgenus *Costus* (Costaceae): evidence from nrDNA ITS and ETS sequences. *American Journal of Botany* 92: 1899–1910.

Kay, Q.O.N. 1965. *Experimental and comparative ecological studies of selected weeds*. Ph.D. Thesis, Univ. Oxford.

Kim, K.J., Choi, K.S. & Jansen, R.K. 2005. Two chloroplast DNA inverstions originated simultaneously during the early evolution of the sunflower family (Asteraceae). *Molecular Biology and Evolution* 22: 1783–1792.

Klak, C., Reeves, G. & Hedderson, T. 2004. Unmatched tempo of evolution in southern African semi-desert ice plants. *Nature* 427: 63-65.

Koch, M.A., Kiefer, C., Ehrich, D., Vogel, J., Brochmann, C. & Mummenhoff, K. 2006. Three times out of Asia Minor: the phylogeography of *Arabis alpina* L. (Brassicaceae). *Molecular Ecology* 15: 825-839.

Kozak, K.H. & Wiens, J.J. 2006. Does niche conservatism promote speciation? A case study in North American salamanders. *Evolution* 60: 2604-2621.

Krajewski, C., Wroe, S. & Westerman, M. 2000. Molecular evidence for the pattern and timing of cladogenesis in dasyurid marsupials. *Zoological Journal of the Linnean Society, London* 130: 375-404.

Kress, W.J., Wurdack, K.J., Zimmer, E.A., Weigt, L.A. & Janzen, D.H. 2005. Use of DNA barcodes to identify flowering plants. *Proceedings of the National Academy of Sciences USA* 102: 8369-8374.

Krijgsman, W. 2002. The Mediterranean: Mare Nostrum of Earth sciences *Earth and Planetary Science Letters* 205: 1-12.

Lambeck, K., Esat, T.M. & Potter, E.K. 2002. Links between climate and sea levels for the past three million years. *Nature* 419: 199-206.

Lavin, M., Wojciechowski, M.F., Gasson, P., Hughes, C. & Wheeler, E. 2003. Phylogeny of robinioid legumes (Fabaceae) revisited: *Coursetia* and *Gliricidia* recircumscribed, and a biogeographical appraisal of the Caribbean endemics. *Systematic Botany* 28: 387 - 409.

Lee, C. & Wen, J. 2004. Phylogeny of *Panax* using chloroplast *trnC-trnD* intergenic region and the utility of *trnC-trnD* in interspecific studies of plants. *Molecular Phylogenetics and Evolution* 31: 894-903.

Linder, H.P. 2003. The radiation of the Cape flora, southern Africa. *Biological Reviews* 78: 597-638.

Linder, H.P. 2006. Investigating the evolution of floras: problems and progress - an introduction. *Diversity & Distributions* 12: 3-5.

Linder, H.P. 2008. Plant species radiations: where, when, why? *Philosophical Transactions of the Royal Society B: Biological Sciences* 363: 3097–3105.

Liston, A. & Wheeler, J.A. 1994. The phylogenetic position of the genus *Astragalus* (Fabaceae): evidence from the chloroplast genes *rpoC1* and *rpoC2*. *Biochemical Systematics and Ecology* 22: 377-388.

Lo Presti, R.M. & Oberprieler, C. 2009. Evolutionary history, biogeography and eco-climatological differentiation of the genus *Anthemis* L. (Compositae, Anthemideae) in the circum-Mediterranean area. *Journal of Biogeography* 36: 1313-1332.

Lo Presti, R.M., Oppolzer, S., Oberprieler, C. 2010. A molecular phylogeny and a revised classification of the Mediterranean genus *Anthemis* s.l. (Compositae, Anthemideae) based on three molecular markers and micromorphological characters. *Taxon* 59: 1441-1456.

Lomolino, M.V. & Heaney, L.R. 2004. Frontiers of biogeography: new directions in the geography of nature. *Sinauer Associates, Sunderland, MA.*

Mansion, G., Selvi, F., Guggisberg, A. & Conti, E. 2009. Origin of Mediterranean insular endemics in the Boraginales: integrative evidence from molecular dating and ancestral area reconstruction. *Journal of Biogeography* 36: 1282-1296.

Mansion, G., Zeltner, L. & Bretagnolle, F. 2005. Phylogenetic patterns and polyploid evolution within the Mediterranean genus *Centaurium* (Gentianaceae - Chironieae). *Taxon* 54: 931-950.

Mantel, N. 1967. The detection of disease clustering and a generalized regression approach. *Cancer Research* 27: 209–220.

Martínez, F. & Montero, G. 2004. The *Pinus pinea* L. woodlands along the coast of southwestern Spain: data for a new geobotanical interpretation. *Plant Ecology* 175: 1-18.

Martini, R., Cirilli, S., Saurer, C., Abate, B., Ferruzza, G. & Lo Cicero, G. 2007. Depositional environment and biofacies characterisation of the Triassic

(Carnian to Rhaetian) carbonate succession of Punta Bassano (Marettimo Island, Sicily). *Facies* 53: 389-400.

Marty, B., Trull, T., Lussiez, P., Basile, I. & Tanguy, J.C. 1994. He, Ar, O, Sr and Nd isotope constraints on the origin and evolution of Mount Etna magmatism. *Earth and Planetary Science Letters* 126: 23-39.

Mason-Gamer, R.J., Holsinger, K.E. & Jansen, R.K. 1995. Chloroplast DNA haplotype variation within and among populations of *Coreopsis grandiflora* (Asteraceae). *Molecular Biology and Evolution* 12: 371-381.

McCauley, D.E., Smith, R.A., Lisenby, J.D. & Hsieh, C. 2003. The hierarchical spatial distribution of chloroplast DNA polymorphism across the introduced range of *Silene vulgaris*. *Molecular Ecology* 12: 3227-3235.

Médail, F. & Diadema, K. 2009. Glacial refugia influence plant diversity patterns in the Mediterranean Basin. *Journal of Biogeography* 36: 1333-1345.

Médail, F. & Quézel, P. 1997. Hot-spots analysis for conservation of plant biodiversity in the Mediterranean Basin. *Annals of the Missouri Botanical Garden* 84: 112-127.

Meister, J., Hubaishan, M., Kilian, N. & Oberprieler, C. 2006. Temporal and spatial diversification of the shrub *Justicia areysiana* Deflers (Acanthaceae) endemic to the monsoon affected coastal mountains of the southern Arabian Peninsula. *Plant Systematics and Evolution* 262: 153-171.

Meulenkamp, J.E. & Sissingh, W. 2000. Tertiary. Pp. 153-208. in: Crasquin, S. (ed.), *Atlas Peri-Tethys, Palaeogeographical Maps - Explanatory Notes.* CCGM/CGMW, Paris.

Meusel, H. & Jager, E. 1992. *Vergleichende Chorologie der zentraleuropaischen Flora.* 3. Gustav Fischer Verlag, Jena.

Miller, J.T. & Bayer, R.J. 2001. Molecular phylogenetics of *Acacia* (Fabaceae: Mimosoideae) based on the chloroplast *matK* coding sequence and flanking *trnK* intron spacer regions. *American Journal of Botany* 88: 697-705.

Mitsuoka, S. & Ehrendorfer, F. 1972. Cytogenetics and evolution of *Matricaria* and related genera (Asteraceae-Anthemideae). *Österreichische botanische Zeitschrift* 120: 155-200.

Mittermeier, R.A., Gil, P.R., Hoffman, M., Pilgrim, J., Brooks, T., Mittermeier, C.G., Lamoreux, J. & da Fonseca, G.A.B. 2005. *Hotspots*

revisited: earth's biologically richest and most threatened terrestrial ecoregions.
Cemex, Conservation International and Agrupacion Sierra Madre, Monterrey,
Mexico.

Moore, A.J. & Bohs, L. 2003. An ITS phylogeny of Balsamorhiza and Wyethia
(Asteraceae: Heliantheae). *Am. J. Bot.* 90: 1653-1660.

Mouterde, P. 1983. *Anthemis* L. Pp. 402-415. in: A. Charpin and Dittrich, M.
(ed.), *Nouvelle Flore du Liban et de la Syrie.* Dar el-Machred Sarl, Beyrouth.

Muller, H. 1883. *The fertilisation of flowers.* Mc Millan & Co, London.

Myers, N. 1988. Threatened biotas: "hot spots" in tropical forests. *The
Environmentalist* 8: 187-208.

**Myers, N., Mittermeier, R.A., Mittermeier, C.G., da Fonseca, G.A.B. &
Kent, J.** 2000. Biodiversity hotspots for conservation priorities. *Nature* 403:
853-858.

Nei, M. & Li, W.H. 1979. Mathematical model for studying genetic variation in
terms of restriction endonucleases. *Proceedings of the National Academy of
Sciences USA* 76: 5269–5273.

Nix, H.A. 1986. A biogeographic analysis of Australian elapid snakes. *Pp. 4-5 in
Atlas of elapid snakes of Australia. Australia Governement Publishing Service,
Canberra, Australia.*

Nixon, K.C. 1999. The parsimony ratchet, a new method for rapid parsimony
analysis. *Cladistics* 15: 407 - 414.

Oberprieler, C. 1998. The systematics of *Anthemis* L. (Compositae,
Anthemideae) in W and C North Africa. *Bocconea* 9: 1-328.

Oberprieler, C. 2001. Phylogenetic relationships in *Anthemis* L. (Compositae,
Anthemideae) based on nrDNA ITS sequence variation. *Taxon* 50: 745-762.

Oberprieler, C. 2004a. On the taxonomic status and the phylogenetic
relationships of some unispecific Mediterranean genera of *Compositae-
Anthemideae* I. *Brocchia, Endopappus,* and *Heliocauta. Willdenowia* 34: 39-57.

Oberprieler, C. 2004b. On the taxonomic status and the phylogenetic
relationships of some unispecific Mediterranean genera of Compositae-
Anthemideae II. *Daveaua, Leucocyclus,* and *Nananthea. Willdenowia* 34: 341-
350.

Oberprieler, C. 2005. Temporal and spatial diversification of Circum-Mediterranean Compositae-Anthemideae. *Taxon* 54: 951-966.

Oberprieler, C., Himmelreich, S. & Vogt, R. 2007. A new subtribal classification of the tribe Anthemideae (Compositae). *Willdenowia* 37: 89-114.

Oberprieler, C. & Vogt, R. 2000. The position of *Castrilanthemum* Vogt & Oberprieler and the phylogeny of Mediterranean *Anthemideae* (*Compositae*) as inferred from nrDNA ITS and cpDNA*trnL/trn*F IGS sequence variation. *Plant Systematics and Evolution* 225: 145-170.

Oberprieler, C. & Vogt, R. 2006. The taxonomic position of *Matricaria macrotis* (Compositae-Anthemideae). *Willdenowia* 36: 329-338.

Oberprieler, C., Watson, L.E. & Vogt, R. 2006 ["2007"]. Tribe *Anthemideae* Cass. Pp. 342-374. in: J. Kadereit and Jeffrey, C. (ed.), *The families and genera of vascular plants.* Springer Verlag.

Oppolzer, S. 2007. *Untersuchungen zu Mikromerkmalen und deren Evolution in Anthemis (Compositae, Anthemideae).* Undergraduation Thesis, University of Regensburg.

Ortiz, M.Á., Tremetsberger, K., Stuessy, T.F., Terrab, A., García-Castaño, J.L. & Talavera, S. 2009. Phylogeographic patterns in *Hypochaeris* section *Hypochaeris* (Asteraceae, Lactuceae) of the western Mediterranean. *Journal of Biogeography* 36: 1384-1397.

Parenti, L. & Humphries, C.J. 2004. Historical biogeography, the natural science. *Taxon* 53: 899-903.

Paun, O., Lehnebach, C., Johansson, J.T., Lockhart, P. & Hörandl, E. 2005. Phylogenetic relationships and biogeography of *Ranunculus* and allied genera (Ranunculaceae) in the Mediterranean region and in the European Alpine System. *Taxon* 54: 911-932.

Peakall, R. & Smouse, P.E. 2006. GENALEX 6: genetic analysis in Excel. Population genetic software for teaching and research. *Molecular Ecology Notes* 6: 288-295.

Pedley, H.M., House, M.R. & Waugh, B. 1976. The geology of Malta and Gozo. *Proceedings of the Geologists' Association* 87: 325-341.

Pelser, P.B., Nordenstam, B., Kadereit, J.W. & Watson, L.E. 2007. An ITS phylogeny of tribe Senecioneae (Asteraceae) and a new delimitation of *Senecio* L. *Taxon* 56: 1077-1104.

Petit, R.J., Brewer, S., Bordàcs, S., Burg, K., Cheddadi, R., Coart, E., Cottrell, J., Csaikl, U.M., van Dam, B., Deans, J.D., Espinel, S., Fineschi, S., Finkeldey, R., Glaz, I., Goicoechea, P.G., Jensen, J.S., König, A.O., Lowe, A.J., Madsen, S.F., Màtyàs, G., Munro, R.C., Popescu, F., Slade, D., Tabbener, H., de Vries, S.G.M., Ziegenhagen, B., de Beaulieu, J.L. & Kremer, A. 2002. Identification of refugia and post-glacial colonisation routes of European white oaks based on chloroplast DNA and fossil pollen evidence. *Forest Ecology and Management* 156: 49-74.

Pfeil, B.E. 2009. The effect of incongruence on molecular dates. *Taxon* 58: 511-518.

Phillips, S.J., Anderson, R.P. & Schapire, R.E. 2006. Maximum entropy modeling of species geographic distributions. *Ecological Modelling* 190: 231-259.

Piñeiro, R., Aguilar, J.F., Munt, D.D. & Feliner, G.N. 2007. Ecology matters: Atlantic-Mediterranean disjunction in the sand-dune shrub Armeria pungens (Plumbaginaceae). *Molecular Ecology* 16: 2155-2171.

Poli, E. 1965. *La vegetazione altomontana dell'Etna*. Flora et Vegetatio Italica. Sondrio.

Pons, O. & Petit, R.J. 1996. Measuring and testing genetic differentiation with ordered versus unordered alleles. *Genetics* 144: 1237–1245.

Provan, J., Soranzoa, N., Wilson, N.J., Goldstein, D.B. & Powell, W. 1999. A low mutation rate for chloroplast microsatellites. *Genetics* 153: 943–947.

Quézel, P. 1985. Definition of the Mediterranean region and the origin of its flora. Pp. 9-24. in: Gomez-Campo, C. (ed.), *Plant conservation in the Mediterranean area* Dr. W. Junk Publishers, Dordrecht.

Quezel, P. & Barbero, M. 1993. Variations climatiques au Sahara et en Afrique sèche depuis le Pliocène: enseignements de la flore et de la végétation actuelles. *Bulletin d'écologie* 24: 191-202.

Quézel, P. & Médail, F. 2003. *Ecologie et biogéographie des forêts du bassin méditerranéen*. Elsevier. Paris.

Quint, M. & Classen-Bockhoff, R. 2004. *Evolution of Bruniaceae: evidence from molecular and morphological studies.* Unpublished Abstract, Recent Floristic Radiations in the Cape Flora meeting, Zurich.

Rambaut, A. & Drummond, A.J. 2003. Tracer vers. 1.3. MCMC Trace File Analyser. Available at: http://evolve.zoo.ox.ac.uk/beast/.

Reznick, D.N. & Ricklefs, R.E. 2009. Darwin's bridge between microevolution and macroevolution. *Nature* 457: 837-842.

Richardson, J.E., Pennington, R.T., Pennington, T.D. & Hollingsworth, P.M. 2001. Rapid diversification of a species-rich genus of Neotropical rain forest trees. *Science* 293: 2242-2245.

Riddle, B.R. & Hafner, D.J. 2004. The past and future roles of phylogeography in historical biogeography. Pp. 93-110. in: Lomolino, M.V., Heaney, L.R. (ed.), *Frontiers of biogeography - New directions in the geography of nature.* Sinauer Associates, Sunderland, Massachussets.

Rittmann, A., Romano, R. & Sturiale, C. 1973. Some considerations on the 1971 Etna eruption and on the tectonophysics of the Mediterranean area. *International Journal of Earth Sciences* 62: 418-430.

Ronikier, M., Costa, A., Fuertes Aguilar, J., Nieto Feliner, G., Küpfer, P. & Mirek, Z. 2008. Phylogeography of *Pulsatilla vernalis* (L.) Mill. (Ranunculaceae): chloroplast DNA reveals two evolutionary lineages across central Europe and Scandinavia. *Journal of Biogeography* 35: 1650-1664.

Ronquist, F. 1996. *DIVA.* vers. 1.1. Computer program and manual available at http://www.ebc.uu.se/systzoo/research/diva/diva.html

Ronquist, F. 1997. Dispersal-vicariance analysis: a new approach to the quantification of historical biogeography. *Systematic Biology* 46: 193-201.

Ronquist, F. & Huelsenbeck, J.P. 2003. MRBAYES 3: Bayesian phylogenetic inference under mixed models. *Bioinformatics* 19: 1572-1574.

Ronquist, F., Huelsenbeck, J.P. & van der Mark, P. 2005. MrBayes vers. 3.1. Manual. Available at: http://mrbayes.csit.fsu.edu/wiki/index.php/Manual.

Rosenzweig, M.L. 1995. *Species diversity in space and time.* Cambridge University Press, Cambridge.

Rourke, J.P. 1972. Taxonomic studies on *Leucospermum* R.Br. *Journal of South African Botany* 8: 1-194.

Saitou, N. & Nei, M. 1987. The neighbour-joining method: a new method for reconstructing phylogenetic tree. *Molecular Biology and Evolution* 4: 406 - 425.

Sanderson, M.J. 2002. Estimating absolute rates of molecular evolution and divergence times: a penalized likelihood approach. *Molecular Biology and Evolution* 19: 101-109.

Sanderson, M.J. 2003. r8s: inferring absolute rates of molecular evolution and divergence times in the absence of a molecular clock. *Bioinformatics* 19: 301-302.

Savolainen, V. & Forest, F. 2005. Species-level phylogenetics from continental biodiversity hotspots. Pp. 17-30. in: F.T. Bakker, L.W. Chathrou, B. Gravendeel and Pelser, P.B. (ed.), *Plant species-level systematics: new perspectives on pattern & process.* A.R.G. Gantner Verlag, Ruggell, Liechtenstein.

Schaffer, W.M. & Gadgil, M.D. 1975. Selection for optimal life histories in plants. Pp. 142-157. in: Cody, M. and Diamond, J. (ed.), *The ecology and evolution of communities.* Harvard University Press, Cambridge.

Schembri, P.J. 2003. Current state of knowledge of the Maltese non-marine fauna. Pp. 33-65. in: (ed.), *Malta Environment and planning authority annual report and accounts 2003.* Floriana, Malta.

Serrato-Cruz, M.A., Silva-Rojas, H.V., Valadez-Moctezuma, E. & Zelaya-Molina, L.X. Phylogeny of *Tagetes* genera in Mexico using nuclear genes. Unpublished.

Shannon, C.E. & Weaver, W. 1949. *The mathematical theory of communication.* Univ. Illinois Press. Urbana.

Simmons, M.P. & Ochoterena, H. 2000. Gaps as characters in sequence-based phylogenetic analyses. *Systematic Biology* 49: 369 - 381.

Simurda, M.C., Marshall, D.C. & Knox, J.S. 2005. Phylogeography of the narrow endemic, *Helenium virginicum* (Asteraceae), based upon ITS sequence comparisons. *Systematic Botany* 30: 887-898.

Smissen, R.D. & Breitwieser, I. 2008. Species relationships and genetic variation in the New Zealand endemic *Leucogenes* (Asteraceae: Gnaphalieae). *Journal of Botany* 46 65-76.

Speranza, F., Villa, I.M., Sagnotti, L., Florindo, F., Cosentino, D., Cipollari, P. & Mattei, M. 2002. Age of the Corsica-Sardinia rotation and Liguro-

Provencal Basin spreading: new paleomagnetic and Ar/Ar evidence. *Tectonophysics* 347: 231–251.

Steane, D.A., Nicolle, D., MacKinnon, G.E., Vaillancourt, R.E. & Potts, B.M. 2002. Higher-level relationships among the eucalypts are resolved by ITS sequence data. *Australian Systematic Botany* 15: 49-62.

Stebbins, G.L. 1942. The genetic approach to problems of rare and endemic species. *Madroño* 6: 241-258.

Stebbins, G.L. 1950. *Variation and evolution in plants*. Columbia University Press, New York.

Stebbins, G.L. 1952. Aridity as a stimulus to plant evolution. *American Naturalist* 86: 33-44.

Stöck, M., Sicilia, A., Belfiore, N., Buckley, D., Lo Brutto, S., Lo Valvo, M. & Arculeo, M. 2008. Post-Messinian evolutionary relationships across the Sicilian channel: Mitochondrial and nuclear markers link a new green toad from Sicily to African relatives. *BMC Evolutionary Biology* 8: 56.

Stockwell, D.R.B. & Peters, D.P. 1999. The GARP modelling system: problems and solutions to automated spatial prediction. *International Journal of Geographic Information System* 13: 143-158.

Street, F.A. & Gasse, F. 1981. Recent developments in research into the Quaternary climatic history of the Sahara. Pp. 8-28. in: Allen, J.A. (ed.), *The Sahara: ecological change and early economic history*. Menas Press, London.

Stuessy, T.F., Jakubowsky, G., Gomez, R.S., Pfosser, M., Schluter, P.M., Fer, T., Byung-Yun, S. & Hidetoshi, K. 2006. Anagenetic evolution in island plants. *Journal of Biogeography* 33: 1259-1265.

Suc, J.P. 1984. Origin and evolution of the Mediterranean vegetation and climate in Europe. *Nature* 307: 429–432.

Swofford, D.L. 2002. *PAUP* Phylogenetic Analysis Using Parsimony (*and other Methods)*. vers. 4.0b10 Sunderland, Massachusetts Sinauer Associates

Tadesse, M. 2004. New species, a new combination and synonyms in Compositae from NE Africa. *Nordic Journal of Botany* 22: 667-671.

Templeton, A.R., Crandall, K.A. & Sing, C.F. 1992. A cladistic analysis of phenotypic associations with haplotypes inferred from restriction endonuclease mapping and DNA sequence data. III. Cladogram estimation. *Genetics* 132: 619-633.

Thiede, J. 1978. A glacial Mediterranean. *Nature* 276: 680-683.

Thompson, J.D. 2005. *Plant evolution in the Mediterranean.* Oxford University Press. Oxford.

Thompson, J.D., Higgins, D.G. & Gibson, T.J. 1994. CLUSTAL W: improving the sensitivity of progressive multiple sequence alignment through sequence weighting, position-specific gap penalties and weight matrix choice. *Nucleic acid research* 22: 4673-4680.

Thompson, J.D., Lavergne, S., Affre, L., Gaudeul, M. & Debussche, M. 2005. Ecological differentiation of Mediterranean endemic plants. *Taxon* 54: 967-976.

Tribsch, A. & Schönswetter, P. 2003. Patterns of endemism and comparative phylogeography confirm palaeo-environmental evidence for Pleistocene refugia in the Eastern Alps. *Taxon* 52: 477-497.

Troia, A., Conte, L. & Cristofolini, G. 1997. Isolation and biodiversity in *Cytisus villosus* Pourret (Fabaceae, Genisteae): enzyme polymorphism in disjunct populations. *Plant Biosystems* 131: 93-101.

Uitz, H. 1970. *Cytologische und bestäubungsexperimentelle Beiträge zur Verwandtschaft und Evolution der Anthemideae (Asteraceae).* Ph.D. Thesis, Univ. Graz.

Van Dam, J.A. 2006. Geographic and temporal patterns in the late Neogene (12-3 Ma) aridification of Europe: the use of small mammals as paleoprecipitation proxies. *Palaeogeography, Palaeoclimatology, Palaeoecology* 238: 190-218.

van der Niet, T. & Johnson, S.D. 2009. Patterns of plant speciation in the Cape floristic region. *Molecular Phylogenetics and Evolution* 51: 85-93.

Vargas, P. 2003. Molecular evidence for multiple diversification patterns of alpine plants in Mediterranean Europe. *Taxon* 52: 463-476.

Véla, E. & Benhouhou, S. 2007. Évaluation d'un nouveau point chaud de biodiversité végétale dans le Bassin méditerranéen (Afrique du Nord). *Comptes Rendus Biologies* 330: 589–605.

Vendramin, G.G., Fady, B., González-Martínez, S.C., Hu, F.S., Scotti, I., Sebastiani, F., Soto, Á., Petit, R.J. & Kohn, J. 2008. Genetically depauperate but widespread: the case of an emblematic Mediterranean pine. *Evolution* 62: 680-688.

Verboom, G.A., Archibald, J.K., Bakker, F.T., Bellstedt, D.U., Conrad, F., Dreyer, L.L., Forest, F., Galley, C., Goldblatt, P., Henning, J.F., Mummenhoff, K., Linder, H.P., Muasya, A.M., Oberlander, K.C., Savolainen, V., Snijman, D.A., Niet, T.v.d. & Nowell, T.L. 2009. Origin and diversification of the Greater Cape flora: Ancient species repository, hot-bed of recent radiation, or both? *Molecular Phylogenetics and Evolution* 51: 44-53.

Verboom, G.A., Linder, H.P. & Stock, W.D. 2003. Phylogenetics of the grass genus *Ehrharta*: evidence for radiation in the summer-arid zone of the South African Cape. *Evolution* 57: 1008 - 1021.

Wagenitz, G. 1968. 43. *Anthemis* L. Pp. 292-309. in: Hegi, G. (ed.), *Illustrierte Flora von Mitteleuropa, ed. 2*. Berlin.

Wagstaff, S.J. & Breitwieser, I. 2002. Phylogenetic relationships of New Zealand Asteraceae inferred from ITS sequences. *Plant Systematics & Evolution* 231: 203-224.

Wagstaff, S.J., Breitwieser, I. & Swenson, U. 2006. Origin and relationships of the austral genus *Abrotanella* (Asteraceae) inferred from DNA sequences. *Taxon* 55: 95-106.

Wallmann, P.C., Mahood, G.A. & Pollard, D.D. 1988. Mechanical models for correlation of ring-fracture eruptions at Pantelleria, Strait of Sicily, with glacial sea-level drawdown. *Bulletin of Volcanology* 50: 327-339.

Waterman, R.J., Pauw, A., Barraclough, T.G. & Savolainen, V. 2009. Pollinators underestimated: a molecular phylogeny reveals widespread floral convergence in oil-secreting orchids (sub-tribe Coryciinae) of the Cape of South Africa. *Molecular Phylogenetics and Evolution* 51: 100-110.

Watson, L.E., Bates, P.L., Evans, T.M., Unwin, M.M. & Estes, J.R. 2002. Molecular phylogeny of Subtribe Artemisiinae (Asteraceae), including *Artemisia* and its allied and segregate genera. *BMC Evolutionary Biology* 2: 17.

Westberg, E. & Kadereit, J.W. 2009. The influence of sea currents, past disruption of gene flow and species biology on the phylogeographical structure of coastal flowering plants. *Journal of Biogeography* 36: 1398-1410.

Wiens, J.J. 1998. Combining data sets with different phylogenetic histories. *Systematic Biology* 47: 568 - 581.

Wiens, J.J., Chippindale, P.T. & Hillis, D.M. 2003. When are phylogenetic analyses misled by convergence? A case study in Texas cave salamanders. *Systematic Biology* 52: 501-514.

Wiens, J.J. & Donoghue, M.J. 2004. Historical biogeography, ecology and species richness. *Trends in Ecology & Evolution* 19: 639-644.

Wikström, N., Savolainen, V. & Chase, M.W. 2001. Evolution of the angiosperms: calibrating the family tree. *Proceedings of the Royal Society B: Biological Sciences* 268: 2211–2220.

Wolfe, J.A. 1997. Relations of environmental change to angiosperm evolution during the late Cretaceous and Tertiary. Pp. 269-290. in: Iwatsuki K. and Raven, P.H. (ed.), *Evolution and diversification of land plants.* Springer, Tokyo Berlin Heidelberg New York.

Wolfe, K.H., Li, W.H. & Sharp, P.M. 1987. Rates of nucleotide substitution vary greatly among plant mitochondrial, chloroplast and nuclear DNA. *Proceedings of the National Academy of Sciences USA* 84: 9054–9058.

Wright, S. 1943. Isolation by distance. *Genetics* 28: 114-138.

Yang, Z. & Rannala, B. 2006. Bayesian estimation of species divergence times under a molecular clock using multiple fossil calibrations with soft bounds. *Molecular Biology and Evolution* 23: 212-226.

Yavin, Z. 1970. A biosystematic study of *Anthemis* section *Maruta* (Compositae). *Israel Journal of Botany* 19: 137-154.

Yavin, Z. 1972. New taxa of *Anthemis* from the Mediterranean and SW Asia. *Israel Journal of Botany* 21: 168-178.

Yesson, C. & Culham, A. 2006. Phyloclimatic modeling: combining phylogenetics and bioclimatic modeling. *Systematic Biology* 55: 785-802.

Yokoyama, Y., Lambeck, K., De Deckker, P., Johnston, P. & Fifield, L.K. 2000. Timing of the Last Glacial Maximum from observed sea-level minima. *Nature* 406: 713-716.

Young, N. & Healy, J. 2003. GapCoder automates the use of indel characters in phylogenetic analysis. *BMC Bioinformatics* 4: 6.

Acknowledgments

First and foremost, I am deeply grateful to my teacher Prof. Dr. Christoph Oberprieler who devoted so much of his time to me and this undertaking. His professional advice and creative ideas as well as his thorough challenging of my results and conclusions have greatly contributed to this work. Throughout these years, he has given me the opportunity to present my results at various occasions and to collect valuable feedback on the ongoing research. Without his enthusiasm for our discipline I would likely not have overcome the Alps, yet alone initiate and finally terminate this thesis.

I am also very grateful to Prof. Dr. Dirk Albach from the University of Oldenburg, who kindly accepted to act as referee of this thesis and without hesitation took on the burden to examine this study and come to Regensburg. Many thanks to Prof. Dr. Tod Stuessy from the University of Vienna for acting as third referee and for providing valuable comments on this work.

Likewise, I wish to express my sincere gratitude to Prof. Dr. Silvio Fici from the University of Palermo for his continuous encouragement during the years of this thesis. He initially introduced me to this area of the natural sciences and paved the way to my first research stay in Berlin and finally to this doctorate.

I owe many thanks to my colleagues Roland Hößl, Dr. Sven Himmelreich and Dr. Jörg Meister; their moral support as well as their assistance ranging from methodological suggestions to the setup of restive software is greatly appreciated. Peter Hummel deserves special thanks for his very dependable assistance in the laboratory. Thanks also to Richard Landstorfer, Tobias Engl and Jennifer Flechsler for support in the laboratory during their practical training.

PD Dr. Christoph Reisch has patiently addressed many series of questions on AFLP topics, while Dr. Erik Welk (Halle) and Sebastian Dötterl have kindly assisted me with many GIS issues. Many thanks also to Stephanie Oppolzer for sharing her data on micromorphology.

I would also like to thank everybody who joined me on the sampling excursions throughout Sicily, the surrounding islands and Tunisia; my father Michele Lo Presti has been a very patient travel fellow on many occasions and is now able to identify my plants without fault. I am much indebted to Dr. Saskia Nentwig, Dr. Kathrin Bylebyl, Dr. Gerriet Fokuhl, Dr. Sven Himmelreich, Dr. Jörg Meister and Dr. Martijn Kos, who were with me in Marettimo and sampled the plants instead of having a swim and to Dr. Agostino D'Amico, Laura Brancazio, Caterina Carini, Dr. Germana Porcasi and Dr. Roberta Melazzo who joined me many times across Sicily. Finally, many thanks to Dr. Arne Saatkamp (Marseille), Dr. Errol Véla (Montpellier), Dr. Robert Vogt (Berlin) and Claudia Gstöttl for giving me the opportunity to extend my base data through their sampling efforts in France, Algeria and Tunisia.

I sincerely thank all members of the Institute of Botany of the University of Regensburg and in particular Prof. Dr. Peter Poschlod, Wioletta Moggert, Heike Haist, Christina Meindl, Juliane Drobnik, Dr. Christine Römermann, Susanne Gewolf, Steffen Heelemann, Dr. Maik Bartelheimer, Petra Schitko, Inge Lauer, Günter Kolb for their kind reception and companionship throughout these years.

Financial support was provided by DAAD, Bayerische Forschungsstiftung, DFG and Synthesys and is gratefully acknowledged.

Finally, I'm deeply indebted to my parents and parents-in-law, Hubertus, Giuseppe, Sophie and Ellen: without their seemingly endless patience and moral support I could have never finished this work.

Appendices

Appendix A1: List of all species belonging to *Anthemis* and *Cota*. For the taxa analysed in this thesis (141, corresponding to 75% of all taxa), accession, area adopted in the DIVA analysis (Chapter 3), sample number of the extracted DNA (held at the University of Regensburg), EMBL for ITS, *psbA-trnH* and *trnC-petN* markers are given. Outgroups used in Chapter 2 and 3 are recorded at the bottom of the list.

Taxon	Accession	DIVA	Sample	EMBL ITS	EMBL psbA	EMBL trnC
Anthemis aaronsohnii Eig	Syria, near Aleppo. 3-V -1931.Leg. M.Zohary.Det. A.Eig (W-1962-00851)	f	A575	FM957655	subm.	subm.
Anthemis abrotanifolia (Willd.) Guss.	Gr, Kreta, Ida-Gebirge, Nida Hochebene, 1400-1600 m, 18.06.1982, Hager 759 (B).	d	A261	FM957656	subm.	subm.
Anthemis abylaea (Font Quer & Maire) Oberpr.	Morocco. Tanger Pen.: Ceuta(Sebta), Djebel Fahies(S Djebel Musa), southern slopes of Djebel Fahies above the track between the road Souk-Tleta-Taghramnent-Ceuta and Ben Younnech, limestone rocks and screes. Alt.550-600m.35 53'N/05 24'W.31.05.1993.leg.R.Vogt12045	b	A688	FM957764	subm.	subm.
Anthemis aciphylla Boiss.	Griechenland, Ostägäische Inseln, Lesbos, Höhen zw. Stipsi u. Vafios (39 8,7'N/26 13,1'E - 39 20,5'N/26 13,1'E), Tal s. Vafios, Wegrand, 17-04-00. Herbarium H.Kalheber nr.00-598. (M-0126582)	i	A691	FM957767	subm.	subm.
Anthemis aciphylla2 Boiss.	Turkey, distr. Izmir, in declivibus montis Bozdag, in vicinitate pagi Bozdag. Alt. 1500 - 1800. m. s.m. 14.VIII 1993. Leg. V. Vasak. (W-2003-07100)	i	A743	FM957715	subm.	subm.
Anthemis adonidifolia Boiss.	Turkey. 7 kmS Kamsili. C5 Adana. offener Pinus nigra ssp. pall. Wald. 1150 m. 8-7-1977. H. Dr. F. Sorger 77-28-1. (W-1992-10219)	i	A576	subm.	subm.	subm.
Anthemis aeolica Lojac.	-				-	-
Anthemis aetnensis Schouw	Si, prov. Catania, Monte Etna, near Rifugio Sapienza, 37° 42'N, 15° 00'E, UTM 33S WB0573, Open community on lava, 1970 m, 22.06.1997, S.L.Jury with D.Uzunov, 17394 (B).	c	A263	FM957659	subm.	subm.
Anthemis alpestris (Hoffmanns. & Link) R. Fern.	Hs, Segovia, La Mata, 1180 m, 21.06.1983, Romero s.n. (SALA 39705, Herb. CHO).	a	A116	AJ312786 / AJ312815	subm.	subm.
Anthemis ammanthus Greuter	He, Kasos, Punta Avlak, 5-20 m, 15.04.1983, Raus s.n. (B).	d	A129	AJ312800 / AJ312829	subm.	subm.
Anthemis ammophila Boiss. & Heldr.	Turkey. Antalya. Sanddünen zwischen Kemer und Antalya. 2-5-1970. I. Bozakman, K. Fitz 441. (W-1972-03411)	i	A577	FM957716	subm.	subm.
Anthemis anthemiformis (Freyn & Sint.) Grierson	Turkey. B6 Sivas. 48 km E Kangal. s.d. Str. Baydigziyat T. Felssteppe. 2200 m. 29-6-1970. Def. Grierson. H. Dr. F. Sorger 70-22-30. (W-1992-8640)	j	A578	FM957717	subm.	subm.
Anthemis arenicola Boiss.	Cilicie, pentes du Yemsurdaba Dagh, 07.1872, Péronin n° 135 (S).	i	A457	FM957689	subm.	subm.
Anthemis argyrophylla (Hal. & Georgiev) Velen.	-				-	-
Anthemis arvensis L. subsp. *sphacelata* (C.Presl) R.Fernandes	Si, Madonie, Piano Farina, N. slopes of Mt. Cenasa, 1300 m, woodland of Fagus, Ilex and Quercus petraea on siliceous substrate, roadside, 09.06.1983, W.Greuter & U.Matthäs, 19942 (B).	abcdfghij	A266	FM957661	subm.	subm.
Anthemis arvensis L. subsp. *incrassata* (Loisel.) Nyman	Hs, Prov. Málaga, Serranía de Ronda, Fahrweg von Tolox zur Sierra de las Nieves, Serpentin, ca. 850 m, 26.05.1992, Vogt 9199 (B).	abcdfghij	A015	AJ312777 / AJ312806	subm.	subm.

Appendix A1: Continued.

Taxon	Accession	DIVA	Sample	EMBL ITS	EMBL psbA	EMBL tmC
Anthemis atropatana Iranshahr	Iran, Prov. Azarbaijan orient.: 23 km NW Tabriz versus Sufian. 1350 m. 5-6-1971. K.H. Rechinger, Iter Iranicum VII, 41110. (W-1984-09117)	g	A579	FM957718	subm.	subm.
Anthemis auriculata Boiss.	Gr, N slope near the Midsea Hill, Argos Plain, N slope shrub, found with Arenaria leptoclados and Crepis foetida subsp. commutata. 18.04.1995, J.M.Shay, M578 (B).	i	A267	FM957662	subm.	subm.
Anthemis austroiranica Rech.f., Aellen & Esfand.	Iran, Semnan prov., Touran Protected Area (SE of Shahrud), 10 km E of Chejam, gravelly plains in Zygophyllum-ass., 1200 m, 24.04.1978, Freitag 14730 (B).	g	A029	AJ312798 / AJ312827	subm.	subm.
Anthemis bornmuelleri Stoj. & Acht.	Israel, Palaestina australis: Jaffa, in siccis arenosis. 20.V 1897. Leg. et det. J. Bornmueller. J. Bornmueller - Iter Syriacum - 1897. No. 873. (WU-036518)	efi	A271	FM957664	subm.	subm.
Anthemis bornmuelleri2 Stoj. & Acht.	Israel, upper Galilee, near Ya'ara, 200 m, 05.04.1989, A. Danin, T.Raus, W.Sauer, S.Brullo, B.Valdes, F.Amich, S.G Gardner, R.C.H.J.van Ham, A.Gambino, F.Axelrod, Battia Pazy, Rivka Nokrian, 53038 (B).	efi	A741	FM957784	subm.	subm.
Anthemis bourgaei Boiss. & Reut.	Spanien, Prov. Jean, Sierre de Cazoria, Im Felsschutt der Straße bei Las Empanadas. 19-8-1969. leg. B.&W. Lippert 9927. (M-0126606)	a	A693	FM957768	subm.	subm.
Anthemis boveana J. Gay	Al, wilaya de Tlemcen, Beni-Saf, cap d'Acra, à environ 6 km à l'ouest de Beni-Saf et 50 km au nord de Tlemcen, env. 20 m, 08.05.1986, A.Dubuis, H.Maurel, R.Rhamoun, 18496 (B).	b	A272	FM957665	-	subm.
Anthemis brachycarpa Eig						
Anthemis brachystephana Bornm. & Gauba	Persia, E Khorasan, 56-60 km N Gonabad inter Mahneh, 34°59'N, 58°51'E, et Ermani, 34°35'N, 58°40'E, 900 m, iter iranicum IX 1975, K.H. Rechinger, 51427 (B).	g	A273	FM957666	subm.	subm.
Anthemis breviradiata Eig	Syria, Syrian Desert. 22 km W of Soukhme. Stony and gravelly hill. 1-5-1933. Leg. A. Eig. M. Zohary. Det. A. Eig. (W-1962-00852)	f	A583	FM957720	subm.	subm.
Anthemis bushehrica A44	Iran, Borazjan Chahkhani: 70 m. 25-4-1972. Hortus Botanicus Ariamehr n. 4138. (W-1986-05735)	g	A584	FM957721	subm.	subm.
Anthemis calcarea Sosn.	Turkey. Grobblockschutt Kalk. A9 Erzurum: Yusuleli-Olur 4 km s Olur - 1000 m. 24-6-1988. Herbarium Max Nydegger 43446. (M-MSB-002288)	j	A274	FM957769	subm.	subm.
Anthemis calcarea2 Sosn.	Türkei, Anatolien, A8 Erzurum, 8 km nördlich Tortum, zwischen Tortum und Tortum Gölü, 1450 m, 24.07.1982, M.Nydegger, 15600 (B).	j	A694	FM957667	subm.	subm.
Anthemis candidissima Spreng.	Azerbaijan, Baku, Kischiy, 30.04.1912, Holmberg n° 262 (S).	gi	A458	FM957770	subm.	subm.
Anthemis candidissima2 Spreng.	Georgia, Caucasus centralis, peripheria urbis Tbilisi, haud procul a lacu Lisi. 600 m. 17-5-1985. det. J. Cuba. (M-0126609)	gi	A695	FM957690	subm.	subm.
Anthemis chia L.	Tu, Antalya, Perges, 50 m, 04.04.1992, Oberprieler B. I-25 & Schober (B).	cdethi	A125	AJ312785 / AJ312814	subm.	subm.
Anthemis chrysantha Gay	-			-	-	-
Anthemis confusa Pomel	Tu, Gouvernorat de Tataouine, Tunisie de Sud, road P19 between Tataouine and Remada, c. 15 km N Bir Thlethine, 33°48.763'N, 10°22.475'E, 400 m, 12.05.1994, R.Vogt, C.Oberprieler, 46-939 (B).	b	A280	FM957670	-	subm.
Anthemis cornucopiae Boiss.	Palestine, north of Wadi-Farah, 200 m, 12.03.1935, Dinsmore 9574 (S).		A459	subm.	subm.	subm.
Anthemis corymbulosa Boiss. & Hausskn.						
Anthemis cotula L.	It, Grosseto, Padule di Punta Ala, 15.06.1992, Romi & Marchetti (HB Siena); Cult. in Hort. Bot. Berol. 103-03-93-10 (B).	abodefghij	A032	AJ312794 / AJ312823	subm.	subm.
Anthemis cretica L.	Ga, Aude, La-Torette-Cabardès, HB Liège; Cult. in Hort. Bot. Berol. 126-16-91-10 (CHO).	abchij	A011	AJ296388 / AJ296423	subm.	subm.
Anthemis cretica L. subsp. *cretica*	Si, Madonie, Quacella, 1846 m, 05.2005, Lo Presti s.n. (Lo Presti).	c	A729	FM957778	subm.	subm.

Appendix A1: Continued.

Taxon	Accession	DIVA	Sample	EMBL ITS	EMBL psbA	EMBL tmC
Anthemis cretica L. subsp. *cretica2*	Si, Madonie, 25.05.2007, Oberprieler 10332 (cho).	c	A731	FM957780	subm.	subm.
Anthemis cretica L. subsp. *anatolica* (Boiss.) Grierson	Turkey, Elmalidag, C2 Antalya. S. exp. Steppenhang. c. 1300 m. 23-6-1967. H. Dr. F. Sorger 67-21-106. (W-1992-8712)	abcfhij	A587	FM957723	subm.	subm.
Anthemis cretica L. subsp. *argaea* (Boiss. & Balansa) Grierson	Greece, Ikaria, in saxosis silic. Montis Atheras, 600-900 m, 06.05.1976, Rechinger 54324 (S).	abcfhij	A485	FM957707	subm.	subm.
Anthemis cretica L. subsp. *carpatica* (Willd.) Grierson	Hs, Lerida, Val de Aran, Pic de Pedescals, 2200-2380 m, 31.08.1988, Vogt 7350 & Prem (B).	abcfhij	A119	AJ312787 / AJ312816	subm.	subm.
Anthemis cretica L. subsp. *gerardiana* (Jord.) Greuter	Fr, dép. Vaucluse, Saumane, collines situées au-dessus du golf, 300 m, 21.05.1997, B.Girerd, 18498 (B).	abcfhij	A284	FM957673	subm.	subm.
Anthemis cretica L. subsp. *sibthorpii* (Griseb.) Govaerts	Gr, Mt. Athos (Agion Oros), upper southern side, 1700-2030 m, 27.07.1979, Strid, Papanicolaou, 15926 (B).	abcfhij	A298	FM957681	subm.	subm.
Anthemis cretica L. subsp. *spruneri* (Boiss. & Heldr.) Govaerts	Greece, Pelopomnes, Nom. Korinthia: Oros Killini, N Hange SW Trikala, kalk. 1600-2000 m. 6-7-1985. Leg. et det. D. Podlech n. 39679. (W-2003-13658)	abcfhij	A644	FM957756	subm.	subm.
Anthemis cuneata Hub.-Mor. & Reese	Türkei, Anatolien, C2 Burdur, Westlich des Haci Osman Dagh, 51 km ob Fethiye (Makri), 100 m, 20.06.1981, M.Nyedegger, 13605 (B).	i	A281	FM957671	subm.	subm.
Anthemis cupariana Nyman	Si, Madonie, 25.05.2007, Oberprieler 10331 (cho).	c	A689	FM957765	subm.	subm.
Anthemis cupariana2 Nyman	Italy, Prov. Palermo, Madonie, Piano Zucchi between Collesano and Polizzi, slopes in thin Quercus forest. 37°53,955'N/13°59,582'E. 1100 m. 25-05-1994.leg.R.Vogt 13934 & Ch.Oberprieler 8239	c	A728	FM957777	subm.	subm.
Anthemis cyrenaica Coss.	Libya, Cyrenaica, 23 km ESE of Al Abyar, in small wadi, 22.02.1983, Thor 3412 (S).	b	A433	FM957713	subm.	subm.
Anthemis davisii Yavin	Turkey, B8 Bitlis: 13 km from Baykan to Bitlis, 1150 m. Shaley screes in gullies. Annual. Ligules white. 18-5-1966. Davis 43155. (W-1976-07825)	j	A591	FM957725	subm.	subm.
Anthemis debilifolia Eig	-				-	-
Anthemis deserticola H. Kraschen. & Popov					-	-
Anthemis desert-syriaci Eig	Iraq. Syr. Desert. 380 km W of Baghdad. foot of hill, sandy soil. 660 m (Jebel Jedra-wa-Jidrain). 1-4-1933. Leg. A. Eig, M.Zohary. Det. A. Eig. (W-1962-00835)	f	A593	FM957726	subm.	subm.
Anthemis didymaea Mout.					-	-
Anthemis edumea Eig	Palestine, Edom, between Ziza and Qatram, 15.04.1929, Eig & Zohary s.n. (S).	f	A462	FM957692	subm.	subm.
Anthemis emaisensis Eig					-	-
Anthemis filicaulis (Boiss. & Heldr.) Greuter	Greece, Creta, Distr. Sitia: in saxosis calc. inter promontorium Sidero et Eremupolis. 5-5-1942. K.H. Rechinger fil.: Iter Aegaeum VI, 1942 n. 12675. (W-1960-11638)	d	A596	FM957728	subm.	subm.
Anthemis fimbriata Boiss.	Turkey, C4 Konya. 63 km N Mut (Str. Karaman). Steppe. c. 1200 m. 11-6-1966. Det. Grierson. H. F. Sorger 66-35-7. (W-1992-10261)	j	A597	FM957729	subm.	subm.
Anthemis freitagii Iranshahr	Afghanistan. Kabul: 2-15 km W Sarobi. 34°40'N, 69°46'E, ad viam versus Kabul. 34°30'N, 69°10'E, substr. calc. 100-1300 m. 13-5-1967. K.H. Rechinger, Iter Orientale 1967 n.34386. (W-1986-05226)	g	A598	FM957730	subm.	subm.
Anthemis fruticulosa M. Bieb.	Russia-Dagestan. In calcareis pr.pag. Lewaschi. 3800 ft. 2 jul.1898. leg. Alexcenko. (M-0126617)	j	A697	FM957772	subm.	subm.
Anthemis fumariifolia Boiss.	Turkey, A4 Ankara: 22 km N Ankara. Steppenhuegel W d. Hauptstr. 900 m. 15-6-1971. H. Dr. F. Sorger 71-6-4. (W-1992-8679)	j	A600	FM957732	subm.	subm.
Anthemis fumarioides Hochst. *Anthemis fungosa* Boiss. & Hausskn.	-					

Appendix A1: Continued.

Taxon	Accession	DVA	Sample	EMBL ITS	EMBL psbA	EMBL tmC
Anthemis gayana Boiss.	Iran, Shahrad - Bustam, 10 km S Zamanabad, 1050 m, 30.04.1975, Rechinger 50770 (B).	g	A126	AJ312799 / AJ312828	subm.	subm.
Anthemis gharbensis Oberpr.	Ma, Rharb, Arbaoua - Moulay Bousselham, 10 m, 24.04.1993, Vogt 10161 & Oberprieler 4609 (B).	b	A122	AJ312778 / AJ312807	subm.	subm.
Anthemis gilanica Bornm. & Gauba	Iran, Kashan (Mooteh protected region), in montibus a Muteh (Mooteh) septentrionem versus, 1950-2000 m, 31.05.1974, Rechinger 46880 (S).	g	A463	FM957693	subm.	subm.
Anthemis gillettii Iranshahr						
Anthemis glaberrima (Rech. f.) Greuter	Greece. Creta. Distr. Kissamos, Ins. Grabusa Agria. In saxosis calc. litor. 20-4-1942. K.H. Rechinger fil. Iter Aegaeum VI n. 12114. (W-1960-11639)	d	A602	FM957733	subm.	-
Anthemis glareosa E. A. Durand & Barratte	Libya, barley fields facing Faculty of Science, Sidi El-Masri, Tripoli, 18.04.1967, Boulos 1667 (S).	b	A467	FM957696	-	subm.
Anthemis gracilis Iranshahr				-	-	-
Anthemis hamrinensis Iranshahr	Iraq, Distr. Baghdad. Inter Tigris flumen et montes Jabal Hamrin 40-50 km ab oppido Samarra orientem versus,ca. 34°15'N, 44°E. In collibus arenosis 10 km a Sheikh Mohammed septentr. versus. 4-5-1957. det. K.H. Rechinger. Itinera Orientalia	f	A605	FM957735	subm.	subm.
Anthemis handel-mazzettii Eig	Iraq, Distr. Basra. Desertum meridionale (Southern desert). Prope stationem viae ferreae Walhaila, 40 km SW Basra, "haswa", 10 m, 25-3-1957. det. K.H. Rechinger. K.H. Rechinger, Itinera Orientalia 1956-57 n. 14305. (W-1967-2158)	f	A606	FM957736	subm.	subm.
Anthemis haussknechtii Boiss. & Reut.	Armenia, 21.06.2002, Oberprieler 10122 (cho).	fgj	A196	FM957658	subm.	subm.
Anthemis hebronica Boiss. & Kotschy	Israel, Judean Mts., Jerusalem, near Hadassa Hospital, 31°45'N, 35°08'E, 30.04.1992, A.Danin, s.n. (B).	fg	A285	FM957674	subm.	subm.
Anthemis hemistephana Boiss.				-	-	-
Anthemis hermonis Eig				-	-	-
Anthemis hirtella C.Winkl.	Uzbekistania, brachia autro-occidentalia jugi Hissarici, orientem versus ab opp. Baisun, ad declivia argilloso-arenosa gypseacca prope pag. Tasch-Kak, 30.04.1930, Botschantzew & Vvedenky n° 6722 (S).	g	A464	FM957694	subm.	subm.
Anthemis homalolepis Eig	Irak, Distr. Baghdad, inter Tigris flumen et montes Jabal Hamrin, 24 km ab oppido Samarra orientem versus, ca. 34°15'N, 44° E, 04.05.1957, Rechinger 9580 (S).	f	A465	FM957695	subm.	subm.
Anthemis hyalina DC.	Lx, zwischen Roedgen und Bahnhof Leudelingen, Kleeacker (Trifolium resupinatum) auf Liaston, massenhaft, mit Eruca sativa, 11.07.1964, Reichling 27218 (B).	fgj	A022	AJ312779 / AJ312808	subm.	subm.
Anthemis hydruntina H. Groves	It, Calàbria, prov. Crotone, Marchesato, ca. 0,5 km NE of Cacouri, Torrente Matassa, 39°13'52''N, 16°47'10''E, 360-500 m, 12.06.1997, R.Vogt, 10018487 (B).	h	A286	FM957675	subm.	subm.
Anthemis indurata Delile	Egypt. Rosette, Huegel Mandur. 26-3-1887. P. Ascherson, Iter aegypticum quartum n. 161. Muesum botanicum Berolinense. (W-1914-9521)	f	A608	FM957737	subm.	subm.
Anthemis ismelia Lojac.	Si, Rocca Busambra, 05.2005, Fici s.n. (Lo Presti).		A732	subm.	subm.	subm.
Anthemis jordanovii Stoy. & Acht.				-	-	-
Anthemis kandaharica Iranshahr	Afghanistan, prov. Helmand: an der Straßengabelung Girishk - Kandahar - Lashkargah, 870 m, halbwüste. 64/49-31/43. 17-4-1978. leg.et det. D.Podlech nr.30693 (M-MSB-002286)	g	A699	FM957773	subm.	subm.
Anthemis kotschyana Boiss.	Türkei, Anatolien, C6 Hatay, Yeditepe-Dagi, zwischen Yayladagi und Antakya, 1260 m, 29.04.1990, M.Nydegger, 18499 (B).	ij	A287	FM957676	subm.	-
Anthemis kruegeriana Pampan.				-	-	-
Anthemis kurdica Iranshahr				-	-	-
Anthemis laconica Franzén	Greece. Pelop., Laconia, Mt. Taygetos, towards Phrol.-Ilias, above Katafigion, 1600 m, 08.06.1996, Emanuelsson 1953 (S).	i	A468	FM957697	subm.	subm.

Appendix A1: Continued.

Taxon	Accession	DIVA	Sample	EMBL ITS	EMBL psbA	EMBL tmC
Anthemis leptophylla Eig	Iraq, Distr. Kirkuk (Kurdistan): 9 km a Kirkuk versus Altun Kopru, inter segetes, 22-4-1957. Leg. G. Erdtman & W.F.C. Goedemans. K.H. Rechinger Itinera Orientalia 1956-57 n. 15510a. (W-1981-09767)	fg	A612	FM957738	subm.	subm.
Anthemis leucanthemifolia Boiss. & Blanche	Israel. Palaestina australis: Jaffa, in siccis arenosis. 7-5-1897. Leg. et det. J. Bornmueller. J. Bornmueller Iter Syriacum 1897 n.869. (W-1898-4346)	f	A613	FM957739	subm.	subm.
Anthemis leucolepis Eig	-				-	-
Anthemis lithuanica (DC.) Besser ex Trautv.	-				-	-
Anthemis lorestanica Iranshahr	Iran. Prov. Lorestan, Khorramabad, Zagheh. Slope: n° 25, soil clay, stony. Flower: white. Height: 15cm. 2000 m. 5-5-1973. Leg. M.Riazi, Hortus Botanicus Anamehr n. 9647. (W-1986-05734)	i	A614	FM957774	subm.	subm.
Anthemis macedonica Boiss. & Orph.	Gr, Macedonia or., prov. Drrama: in ditione Elata (Kara Dere), a pago Skaloti 10 km septemtriones versus, 41°29′N, 24°18′E, 1550 m, 19.08.1978, W.Greuter, 16573 (B).	i	A289	FM957677	subm.	subm.
Anthemis macedonica Boiss. & Orph. subsp. *orbelica* (Pančić) Oberpr. & Greuter	He, Drama, Zagradenia - Frakto, 1540 m, 13.07.1987, Oberprieler 2992 (B).	i	A117	AJ312790 / AJ312819	subm.	subm.
Anthemis macrotis (Rech. f.) Oberpr. & Vogt	He, Saria, Hauptgipfel der Insel («Monte Grosso», Pachivouno), 22.05.1983, Raus 8360 (B).	i	A186	AMf 76761	subm.	subm.
Anthemis maris-mortui Eig	-				-	-
Anthemis maritima L.	Tn, Bizerte, Cap Bizerte, 20 m, 22.05.1994, Vogt 13813 & Oberprieler 8118 (B).	ab	A012	AJ312788 / AJ312817	subm.	subm.
Anthemis maritima L.	Hs, Cabo Roche NW Conil de la Frontera, pine forest around the lighthouse, 5-10 m, 17.04.1993, Vogt 9733 & Oberprieler 4181 (B).	ab	A124	AJ312789 / AJ312818	subm.	subm.
Anthemis marrocana Batt. & Pitard	Morocco, Middle Atlas, Massif du Kandar S Sefrou, road P20 between Sefrou and Boulmane, 33°45′N, 04°51′E, 1150 m, 28.04.1993, R.Vogt, C.Oberprieler, 38-596 (B).	b	A290	FM957678	subm.	subm.
Anthemis marschalliana Willd.	Russland, ...14.06.1964, Andrejev & Ponomareva s.n. (B).	j	A276	FM957668	subm.	subm.
Anthemis marschalliana 2 Willd.	Soviet Union, Kabardino-Blakarskaya ASSR, Bolshoy Kavzkaz, Elbrus area, Cheget mountain, 3040 m, 21.06.1982, Klackenberg 820621-64 (S).	j	A471	FM957698	subm.	subm.
Anthemis mauritiana Maire & Sennen	Morocco, Mediterranean coast W of Sadia, surroundings of the mouth of Oued Moulouya c.10 km W Saidia, dunes c. 1km inland, 35°07′N, 02°18′E, 10 m, 05.05.1993, R.Vogt, Ch.Oberprieler, 64-1171 (B).	b	A291	FM957679	subm.	subm.
Anthemis melampodina Delile	Sn, Sinai, Peninsula, 20 km W of Nizzana at road to El Aresh, 130 m, sandy soils, 34°14′E-30°31′N, 02.05.1991, Podlech 49968 (MSB); Cult. in Hort. Bot. Berol. 049-29-93-10, 15.06.1993 (B).	f	A025	AJ312780 / AJ312809	subm.	subm.
Anthemis melampodina Delile subsp. *melampodina*	Kuwait, Al-Khiran, near the Gulf, north of Al-Nuwaisib, frontier station with Saudi Arabia, 24.04.1987, Nilsson 16448, Halwagy & Boulos (S).	f	A461	FM957691	subm.	subm.
Anthemis melanacme Boiss. & Hausskn.	Iraq, 14 13 16 6 Kersi. grassy hillside, white ray fl., yellow tube fl.dispersed. 500 m. Anders 544. (W-1981-09768)	fij	A620	FM957742	subm.	subm.
Anthemis meteorica Hausskn.	Greece, Thessalia, in collius schist. (Grüschieler) a Volos occidentem versus, 03.05.1961, Rechinger 22598 (S).	i	A474	FM957699	subm.	subm.
Anthemis micrantha Boiss. & Hausskn.	-				-	-
Anthemis microcephala (Schrenk) B. Fedtsch.	Iran. Fars: Zarghan near Shiraz. 29-5-1968. Leg. Sharale. Det. Iranshahr (1976). EVIN n. 13192 E. (W-1969-9522)	fg	A623	FM957744	subm.	subm.
Anthemis microlepis Eig	Iraq, Jiddala. grassy slope, S side Jabal Sinjar. dispersal. 1000 m. 11-5-1968. Anders 1976. (W-1970-1875)	f	A624	FM957745	subm.	subm.

Appendix A1: Continued.

Taxon	Accession	DIVA	Sample	EMBL ITS	EMBL psbA	EMBL trnC
Anthemis microsperma Boiss. & Kotschy	Eg. Cairo-Alexandria desert road 180 km from Cairo, 16.04.1967, Täckholm s.n. (S).	f	A476	FM957700	subm.	subm.
Anthemis mirheydari Iranshahr	Iran. Prov. Bandar-Abbas: 22 km N of Qotbad, N side of tunnel. Limestone, 900 m, 10-4-1975. Leg. P. Wendelbo & H. Foroughi, Ariamehr Botanical Garden n. 15820. (W-1975-21811)	g	A626	FM957746	subm.	subm.
Anthemis moghanica Iranshahr	Iran. Azerbaijan: Moghan, Aghran. 12-5-1966. Leg. Soltani et Kashkouli. EVIN n. 6859 E. (W-1966-20218)	g	A627	FM957747	-	-
Anthemis moniliicostata Pomel.						
Anthemis muricata (DC.) Guss.	Si. Località Torrente Vaccarizzo (S. Caterina Villarmosa - prov. Caltanissetta). 460 m ca., 02.04.2004, Lo Presti s.n. (Lo Presti)	c	A746	FM957788	subm.	subm.
Anthemis nabataea Eig	Jordanien. 500 m SW Qasr Amra, Wadi Butm. Ammon. bachbegleitender Steppenwald. 600 m. 17-4-1998. Leg. et det. M. Staudinger. H. M. Staudinger n. J10/27. (W-2005-10889)	f	A630	FM957749	subm.	subm.
Anthemis odontostephana Boiss.	Iran, Khorasan, Kashmar - Rirasth, 1300-1400 m, 04.05.1975, Rechinger 51166 (B).	b	A127	AJ312797 / AJ312826	subm.	subm.
Anthemis orientalis (L.) Degen	Gr. Peloponnes, Nomos Achaia, Eparchia Kalavrita: Chelmos Gebirge (Aroania Ori), 1500 m, 05.1960, F.Sorger, 60-3-4-11 (B).	i	A279	FM957669	subm.	subm.
Anthemis parnasia Boiss. & Heldr.	Griechenland: Flora Attica: in m. Parnethe pr. Dekeleiam (hod. Tatoi). Ad cacumen montis. Mai 1900, leg. Dr.Th.de Heldreich. (M01-26637)	i	A701	FM957775	-	-
Anthemis parvifolia Eig						
Anthemis patentissima Eig						
Anthemis paucibra Boiss.	Turkey. 30 km N Konya (Str. Afyon). Felspartien - ausgetrocknetenes Bachbett. c. 1300 m. 11-06-1966. Det. Grierson. Herbarium Friederike Sorger n. 66-40-23. (W-1992-10230)	f	A478	FM957701	subm.	subm.
Anthemis paucilobaz Boiss.	Iraq, Mosul, ad confines Tuciae prov. Hakari, in ditione pagi Sharansieh, in montibus calc. A Zakho septentrionem versus, in saxosis cacumen Zawita, 2000 m. 4.-9.07.1957, Rechinger 10968 (S).	f	A634	FM957750	subm.	subm.
Anthemis pedunculata Desf.	Ma, Monts des Beni Snassen, Tanezzert - Taforalt, 900-950 m, 09.05.1993, Vogt 11448 & Oberprieler 5896 (B).	ab	A014	AJ312791 / AJ312820	subm.	subm.
Anthemis pedunculata2 Desf.	Ma, Prov. de Taza, Gegend um Taffert an der Straße 4803 von Ahermoumou (Ribat-el-Kheyr) zum Jebel Bou Iblane, Kalkfelsen, 1680 m, 4°15'W - 33°40'N, 25.06.1989, Oberprieler 1908 (CHO); Cult. in Hort. Bot. Berol. 071-43-91-10, 15.05.1992 (B).	ab	A123	AJ312792 / AJ312821	subm.	subm.
Anthemis peregrina L.	Gr, Peloponnes, W-Küste, Nm. Messinias, W Kalamata, Strand zwischen Navarino und Petrochori, 16.04.1987, Lakeberg 3281 (Herb. Lange).	hi	A017	AJ312781 / AJ312810	-	-
Anthemis persica Boiss.	Iran. Gypshügel bei Kasrun. STAPF - Iter persicum no. 1332. (WU-036499)	g	A740	FM957783	subm.	subm.
Anthemis pindicola Halácsy						
Anthemis plebeia Boiss. & Noe					-	-
Anthemis plutonia Meikle	Cy, Stavrovouni, Böschung an der Straße bei c. 500 m, Vulkan-Gestein, 04.1991, Vogt 9009; Cult. in Hort. Bot. Berol. 171-15-9-1-20, 15.05.1992 (B).	e	A013	AJ312793 / AJ312822	subm.	subm.
Anthemis pseudocotula Boiss.	Cy, Paphos, Nata - Axylon, 350-400 m, 22.04.1991, Vogt 8568 (B).	defgij	A128	AJ312795 / AJ312824	subm.	subm.
Anthemis punctata Vahl	Tn, El Kef, Dorsale Tunisienne, Table de Jugurtha, limestone cliffs, stony slopes, 1200-1270 m, 35°44.804'N, 08°22.666'E, 05.05.1994, Vogt 12466 & Oberprieler 6771 (cho, B).	b	A730	FM957779	subm.	subm.
Anthemis pungens Yavin					-	-
Anthemis rascheyana Boiss.	Syria, Coelosysien, Yaat i stepp, 1000 m, 30.05.1932, Wall 543 (S).	f	A480	FM957702	subm.	subm.
Anthemis regis-borisii Stoy. & Acht.					-	-
Anthemis retusa Delile	Eg. Giza, bank of the Nile, 17.02.1927, Täckholm s.n. (S).	f	A481	FM957703	subm.	subm.
Anthemis rhodensis Boiss.					-	-

Appendix A1: Continued.

Taxon	Accession	DIVA	Sample	EMBL ITS	EMBL psbA	EMBL trnC
Anthemis rhodocentra Iranshahr	Persia, Balúchestán, in planitie semideserta in vicinitate pagi Deh Pábid (=Dehpapid), 1420 m, 28°37'-60°46', 28.03.1973, Sojak 385 (S).	g	A482	FM957704	subm.	subm.
Anthemis rigida Heldr.	Cy, Larnaka, Cap Greco, Korallenkalke (Unteres Miozän) an der Küste, 5-20 m, 12.04.1991, Vogt 8044 (B).	dei	A020	AJ312782 / AJ312811	subm.	-
Anthemis rosea Sm.						
Anthemis rumelica (Velen.) Stoj. & Acht.	Greece. Macedonia occidentalis, distr. Phlorina: in jugo inter Kastoria et lacum Presba. c.1000 m. K.H. Rechinger Iter Graecum XVI 1972 n. 44887. (W-1978-17525)	i	A639	FM957752	subm.	subm.
Anthemis ruthenica M. Bieb.	Hungary, prov. Bács-Kiskun, Kecskemét area, Fülöphaza (20 km W of Kecskemét), 150 m, 26.05.1987, Bergqvist & al. 4 (S).	hij	A484	FM957706	subm.	subm.
Anthemis scariosa Banks & Sol.	Syria, Job. Karra, 900 m, 16.04.1938, Dinsmore 18347 (S).	igj	A487	FM957709	subm.	subm.
Anthemis schizostephana Boiss. & Hausskn.	Iraq, Distr. Sulaimaniya (Kurdistan). In ditione pagi Penjwin. In glareosis supen. pagi Malakawa. 1400 m. 19-6-1957. K.H. Rechinger Itinera Orientalia 1956-57 n. 12322. (W-1981-09755)	fg	A640	FM957753	subm.	subm.
Anthemis scopulorum Rech. f.	Greece. Regio Aegaea austr. insula Karpathos: in mendionali scopulorum ad ostia sinus Tristomo. in rupestribus sublitoreis. hoc loco copiose. 5 m. 3-6-1963. W. Greuter iter graecum VI anni 1963 n. 5787. (W-1974-10812)	i	A641	FM957754	subm.	subm.
Anthemis scrobicularis Yavin	Jordanien. 12 km NNO Aqaba, Wadi Labanan, Edom. Chamaephytenflur auf Granitschutthalde und in Wadibett. 860 m. 27.4-1998. Leg. et det. M. Staudinger. H. M. Staudinger n. J26/33. (W-1967-25534)	f	A642	FM957755	subm.	subm.
Anthemis secundiramea Biv.	Sicilia, prov. Trapani, Castelluzzo, 0 m, 30-04-2003, Rosa Maria Lo Presti	d	A308	FM957688	subm.	subm.
Anthemis secundiramea Biv. subsp. urvilleana (DC.) R. Fern.	Me, N coast, Salina Bay, Qawra, on flat calcareous coastal rocks N of Suncrest Hotel, in small and shallow soil accumulations, c. 10 m, 35°57'N - 14°26'E, 24.03.1993, Thiede 2227 (B).	abch	A019	AJ312783 / AJ312812	subm.	subm.
Anthemis sintenisii Freyn	Paphlagonia, Wilajet Kastambuli, Tossia, in collibus ad Szuluk-tschesme, 21.05.1892, Sintenis 3908 (S).	i	A488	FM957710	subm.	subm.
Anthemis spec. (from Cyprus)	Nord-Zypern, Kormacit Burnu/Cape Kormakitis, Felsflur um den Leuchtturm am Kap, Kalk, ca. 2-5m. 28.6.2005. leg. R. Vogt 16276. (B)	e	A566	FM957789	subm.	subm.
Anthemis sterilis Steven	Ukraine: Plantae Tauriae - distr. Sudak, montes Karadagh in rupibus. 400 m. 1-6-1975.leg. et det. A.K.Skvortsov. (M-0126639)	j	A702	FM957776	subm.	subm.
Anthemis stiparum Pomel	-					
Anthemis susiana Nábélek	Iraq, Distr. Amara (Mesopotamia). In collibus arenaceis argillosis conglomeraticis ad stationem munitam Shatt at-Tib, ab Amara c. 70 km sept. versus. 32°30'N, 47°10'E. 50-200 m. 27/28-3-1957. K.H. Rechinger Itinera Orientalia 1956-57 n. 8879. (W-1967-2200)	fg	A645	FM957757	subm.	subm.
Anthemis tauberii Durand & Barratte	-					
Anthemis tenuicarpa Eig						
Anthemis tenuisecta Ball	Morocco, prov. Agadir, road P8 between Agadir and Essaouira, c. 10 km N Tamri, 30°49'N, 09°50'E, 120 m, 22.05.1993, Vogt & Oberprieler 12-2224 (B).	b	A300	FM957682	subm.	subm.
Anthemis tigrensis J.Gay ex A.Rich.	Ethiopia, Mussolini-Pass, Debre Berkan - Debre Sina, 2800-3000 m, 23.07.1965, de Wilde 7368 & de Wilde-Duyfjes (B).	k	A120	AJ312796 / AJ312825	subm.	subm.
Anthemis tomentella Greuter	Gr. Kreta, Eparchia Ierapetras, 1,5 km W Straße Kavousi-Tholos, Obersite des Klipperzuges (Aufnahme 5), 35°08'33''N, 25°50'40''E, UTM 35S LU9849, MTB 0821/3A1, 200m, 14.04.1992, Jahn s.n. (B).	d	A301	FM957683	subm.	subm.
Anthemis tomentosa L.	Turkey, Tekirdag; Pasaalan Deresi, Meeresnaehe, 5-5/8-6-1968. Det. A. Gilli, K. Bauer, K. Fitz, F. Spitzenberger 1968, Pflanzen aus der Tuerkei n. 2823. (W-1995-07769)	i	A648	FM957759	subm.	subm.
Anthemis transcheliana Fed.						-
Anthemis tricolor Boiss.	Cyprus. M. Troodos, in pineto, pineta ""Military Camp"". 22-6-1939. Leg. et det. H. Lindberg. H. Lindberg Iter 1939. (W-194-9-6543)	e	A649	FM957760	subm.	subm.

Appendix A1: Continued.

Taxon	Accession	DIVA	Sample	EMBL ITS	EMBL psbA	EMBL tmC
Anthemis tricornis Eig	Turkey, Prov. Diyarbakir. Meden-Ergani. Igneous slopes. Annual. Ligules white. 1000 m. 2-6-1957. Davis & Hedge (D. 29063). (W-1976-07304)	j	A650	FM957761	subm.	subm.
Anthemis tripolitana Boiss. & Blanche	Lebanon. Ad Berytum (Beirut), in arenosis maritimis ad Ras Beirut. 4 et 8-51910. Leg. J. et F. Bornmueller. J. Bornmueller: Iter Syriacum II (1910) - Iter orientale XI - n. 12003. (W-1912-4098)		A492	subm.	subm.	subm.
Anthemis trotzkiana Bunge	Russland, prov. Volgograd. districtus Zhirnovsk. 12.06.1974, Dronova s.n. (B).	j	A303	FM957684	subm.	subm.
Anthemis tubicina Boiss. & Hausskn.	Iran. Prov. Fars. Yasui-Sisakht. 2000 m. 28-4-1972. Leg. H. Foroughi. Hortus Botanicus Ariamehr n. 4167. (W-1984-09947)	fg	A652	FM957762	subm.	subm.
Anthemis ubensis Pomel	Tunisia. Gouvernorat de El Kef, 36°12.868'N, 08°44637'E, 1000 m, 04.05.1994, Vogt & Oberprieler 11284 (B).	b	A304	FM957685	subm.	subm.
Anthemis virescens Velen.	-			-	-	-
Anthemis wallii Huber-Mor. & Reese	-			-	-	-
Anthemis werneri Stoj. & Acht.	Gr. Kykladen, Palea Kaimeni: Inselmitte, 36°23'40''N, 25°23'30''E, 40-100 m, 05.04.1987, Raus 12400 (B).	i	A305	FM957686	subm.	subm.
Anthemis wettsteiniana Hand.-Mazz.	Turkey, Thracia: in arenosis maritimis propontides prope Radosto. 19-6-1890. Dr. A. de Degen Iter orientale a. 1890. (W-1967-2207)	fg	A653	FM957763	subm.	subm.
Anthemis xylopoda O. Schwarz	-			-	-	-
Anthemis yemensis Podlech	Nord Yemen (Jemen Arabic Republic): Innerjemenitisches Hochland, 29kmN Sana'a an der Straße nach Airam, S'Hänge des Djebel Djerban,2550 m;kalk.3.10.1981 leg.D.Podlech.nr.36238. im Bot. Garten M weiterkultiviert.	k	A690	FM957766	subm.	subm.
Anthemis yemensis2 Podlech	Yemen, about 5 km from the village Sanaban on the track to Jabal Isbil, c. 14°29'N, 44°39'E, 2300 m, 11.04.1997, Kilian 4872 & al. (S).	k	A491	FM957712	subm.	subm.
Anthemis zaiarica Oberpr.	Ma. Khemisset, Zaian, Djebel Tougroulmes, 1310 m, 12.05.1995, Vogt 14840 & Oberprieler 9149 (B).	b	A121	AJ312784 / AJ312813	subm.	subm.
Anthemis zoharyana Eig	Negev, Revicim, sandy loess, 02.04.1952, D' Angelis & Chazaelith n° 578 (S).	f	A495	FM957714	subm.	subm.
Cota abagensis Fed.	-			-	-	-
Cota altissima (L.) J. Gay	Persia, N: Gorgan (Mohammad Reza Shah National Park): Tang-e Rah, ad versuras, 400 m, iter iranicum IX, 1975, K.H.Rechinger, 52491 (B).	adfghij	A264	FM957660	subm.	subm.
Cota amblyolepis (Eig) Holub	Cypro. Ad maris lidus prope Larnaka. 20. VI 1880. Sintenis et Rigo Iter Cyprium 1880. no. 804. (WU-035730)	efj	A736	FM957781	subm.	subm.
Cota antitaurica (Grierson) Holub	-			-	-	-
Cota austriaca (Jacq.) Sch. Bip.	De. Bayern, Regierungsbezirk Oberpfalz, Landkreis Regensburg; am Hirmesberg nördlich Kallmünz, Fränkischer Jura, MTB 6837/2, UTM PV4, ca. 450 m, 22.06.1994, F. Schuhwerk, 94/271 (B).	bfghij	A270	FM957663	subm.	subm.
Cota brachmannii (Boiss. & Heldr.) Boiss.	-			-	-	-
Cota brevicuspis (Bom.) Holub	Iraq. N. Jebel Sindjar. c. 800 m. 27-4-1933. Leg. A. Eig. M. Zohary. Det. A. Eig. (W-1962-00832)	fgj	A581	FM957719	subm.	subm.
Cota coelopoda (Boiss.) Boiss.	Turkey. A3 Bolu. 26 km W Gerede. Sumphwiesen, Bachrand. Feld, Gebuesch. 1100 m. 13-6-1971. H. Dr. F. Sorger 71-2-3. (W-1992-8636)	fgij	A585	FM957722	subm.	subm.
Cota dalmatica (Scheele) Oberpr. & Greuter	Croatia. Dalmatien. Golf von Cattaro. zwischen Weinbergen bei Ober-Stulivo. 11-7-1927. Leg. Dr. E. Korb. H. Dr. E. Korb. (W-1956-2365)	h	A589	FM957724	subm.	subm.
Cota dipsacea (Bornm.) Oberpr. & Greuter	Turkey. B2 Izmir.Boz Dag. Bergsteppe.1300-1500 m. 5-8-1968. H. F. Sorger 68-16-27. (W-1992-10262)	i	A594	FM957727	subm.	subm.
Cota dubia (Steven) Holub	Ukraine: Eupatoria.In arenosis littoris Ponti Euxini. 14-7-1900. nr.630. det. v. Halácsy. (M-0126615)	j	A696	FM957771	subm.	subm.

Appendix A1: Continued.

Taxon	Accession	DIVA	Sample	EMBL ITS	EMBL psbA	EMBL tmC
Cota dumetorum (Sosn.) J.Holub	Caucaso, montane Wiese oberhalb der Touristenstation "Allbek", ca 1800 m, 13.07.1977, Ch.Beurton, 478/76 (B).	acfghij	A283	FM957672	subm.	subm.
Cota fulvida (Grierson) Holub	Turkey. Phrygia: Sultandagh, in jugis alpinis supra Engeli. 1850 m. 28-6-1899. Leg. et det. J. Bornmueller. J. Bornmueller. Iter Anatolicum Tertium 1899 n. 4656. (W-1900-1950)	j	A599	FM957731	subm.	subm.
Cota halophila (Boiss. & Balansa) Oberpr. & Greuter	Syria. Alexandrette, Grande Faille (vagin de la vierge). Terrain calcaire. 50-100 m. 4-1932.Leg. Delbes. (W-1962-00843)	fij	A604	FM957734	subm.	subm.
Cota jailensis (Zefir.) Holub	-		A700	-	-	-
Cota linczevskyi (Fedor.) Lo Presti & Oberprieler	Turkmenistan. S-W Kopetdag, Kara-Kalinskii Gebiet. [...]Berg Akbulak.1200 m. 6-7-1974. leg. Nikitin-Ivanov. det. Nikitin. (M-0126622)	g	A700	FM957774	subm.	subm.
Cota lyonnetioides Boiss. & Kotschy			A616	FM957740	-	subm.
Cota macrantha (Heuff.) Boiss.	Bosnia. In monte Hodza prope Hambuliu. 1370 m. 14-7-1911. leg. K. Maly. (W-1924-9276)	i	A616	FM957740	subm.	subm.
Cota macroglossa (Somm. & Lev.) Lo Presti & Oberprieler	Georgien, Provinz Chewi, Großer Kaukasus: Aufstieg vom Tal des Baches Snos-tskali in Richtung des Dorfes Jul' a ca.12 km SE vom Dorf Kazbegi.1900-2100m. 44 45 E 42 35 N.24.VII 97.Leg. et det. G.M. Schneeweiß.Iter Georgicum 1997 Inst. Bot. Univ. Vindob. (W	fij	A745	FM957787	subm.	subm.
Cota maris-nigri Fed.			A618	-	-	-
Cota mazandaranica (Iranshahr) Lo Presti & Oberprieler	Iran. Mazandaran: Kalardasht (Faret). 1350-1500 m. 23-9-1970. Leg. Termeh. Det.Iranshahr. (W-1984-09720)	g	A618	FM957741	subm.	subm.
Cota markhotensis Fed.			A621	FM957743	subm.	subm.
Cota melancloma (Trautv.) Holub	Turkey. B6 Sivas: Yildizdag, offener Eichenbuschwald Felsfluren. 2000 m. 15-7-1969. H. Dr. F.Sorger 69-58-25. (W-1992-8556)	j	A621	FM957743	subm.	subm.
Cota monantha (Willd.) Oberpr. & Greuter	Georgia. Caucasus occidentalis: distr. Gudauta, in vicinitate pagi Blabuchkhva. 50-150 m. 1-8-1980. Leg. V. Vasak. V. Vasak Iter Caucasicum Secundum 1980. (W-1986-00195)	j	A628	FM957748	subm.	subm.
Cota oretana (Carretero) Oberprieler & Greuter			A739	-	-	-
Cota oxylepis Boiss.	Turkey. Grillek Tape. 1100 m. W. Siehe's botanische Reise nach Cilicien 1895/96. no. 425 - 1896. (WU-036446)	ij	A739	FM957782	subm.	subm.
Cota palaestina Kotschy	Libanon, in reg. Arz Libanon, inter Ejbeha at Enden, in agro lapidoso, 1250 m, 11.06.1933, Samuelsson 5905 (S).	defij	A035	FM957657	subm.	subm.
Cota palaestina2 Kotschy	Cy, Paphos, zwischen Nata und Axylon, Garrigue, Kalk und Kalkmergel, 350-400 m, 22.04.1991, Vogt 8569 (B).	defij	A472	AJ312804 / AJ312833	subm.	subm.
Cota pestalozzae (Boiss.) Boiss.	Turkey. C3 Antalya.9 km S Akseki. Felsblockhalde. 1050-1525 m. 29-5-1962. Det. K. Fitz. H. Dr. F.Sorger 62-57-1. (W-1992-8577)	i	A637	FM957751	subm.	subm.
Cota pestalozzae2 (Boiss.) Boiss.	Turkey. WNW of Dag, c. 40 km NNW of Antalya and c. 25 km WSW of Bucak, 25.04.1990, Dinretz & al. S.n. (S).	i	A483	FM957705	subm.	subm.
Cota rigescens (Willd.) Holub	Georgien, Kleiner Kaukasus, Bakuriani, Gora Kochta, 27.08.1997, Beinhauer & Klemm (B).	j	A130	AJ312801 / AJ312830	subm.	subm.
Cota saguramica (Sosn.) Lo Presti & Oberprieler	Georgien. Prov. et Distr. Tiflis, Saghuramio. 27.VI 1923. Leg. B. Schischkim (?). (W-1958-24326)	cfghij	A744	FM957786	subm.	subm.
Cota samuelssonii (Rech. f.) Oberpr. & Greuter	Syrie. Champs a proximité de la ville de Homs, 11.04.1936, s. coll (S).	F	A486	FM957708	subm.	-
Cota schischkiniana Fed.			A296	-	-	-
Cota segetalis (Ten.) Holub	It, Calabria, prov. Cosenza, Sila Greca, ca. 4-7 km N of Bocchigliero; along road from M. Basilicó to Torre Fidile, 39°26′55′′;27°08′′N, 16°45′02′′-35′′E, 400-700 m, 13.06.1997, R.Vogt, 15564 Herbarium Robert Vogt.	hi	A296	FM957680	subm.	subm.

Appendix A1: Continued.

Taxon	Accession	DIVA	Sample	EMBL ITS	EMBL psbA	EMBL trnC
Cota talyschensis (Fedor.) Lo Presti & Oberprieler	Persia, Prov. Azerbaijan orient., in alveo 15 km E Khalkhal, 2050 m, 16.07.1971, Rechinger 43501 (S).	acfghij	A490	FM957711	subm.	subm.
Cota tinctoria (L.) J. Gay	It., Prov. Pescara, Abruzzese, road between Passo di San Leonardo and Caramanico, c. 9 km SE Caramanico, roadsides, 950-1020 m, 42°07.841'N - 14°02.466'E, 29.05.1994, Vogt 14055 & Oberprieler 8360 (B).	dfghij	A036	AJ312802 / AJ312831	subm.	subm.
Cota tinctoria J.Gay subsp. *subtinctoria* (Dobroc.) Holub	Georgia, Trialetisch Gebirge, c. 2km W Kojori, 44°41'E 41°40'30"N, c. 1000 m, 13-7-1997, leg. M Staudinger, det. GUS, Herb. M. Staudinger, Iter Georgicum 1997 Instituti Botanici Universitatis Vindobon. nr. G2/9. (W-2005-02984).	dfghij	A647	FM957758	subm.	subm.
Cota triumfettii (L.) J. Gay	It., Prov. Palermo, Madonie, road between Collesano and Polizzi, c. 7,5 km N Polizzi, road embankments, 1180 m, 37°50.489'N - 14°00.667'E, 26.05.1994, Vogt 13958 & Oberprieler 8263 (B).	acfghij	A037	AJ312803 / AJ312832	subm.	subm.
Cota wiedemanniana (Fisch. & C. A. Mey.) Holub	Gr., Larisa, Eiassonos, 3,6 km NNO Loutro, Getreidefeld, 39°59'30"N, 21°57'E, 875 m, 24.05.1993, Willing 28050 (B).	ii	A306	FM957687	subm.	subm.

Outgroups

Taxon	Accession	DIVA	EMBL ITS	EMBL psbA	EMBL trnC	Reference
Achillea millefolium L.	-		AY603186	-		Guo & al., 2004
Achillea tenuifolia Lam.	Armenia, 18.06.2002, Oberprieler 10094 (cho).		AJ748762	subm.	subm.	Oberprieler 2004a
Anacyclus clavatus (Desf.) Pers.	-		AJ3296389 / AJ3296424	-	-	Oberprieler & Vogt, 2000
Artemisia vulgaris L.			AF422114			Wagstaff & Breitwieser, 2002
Calendula officinalis L.	-		AM774451	-		Himmelreich & al., 2008
Eumorphia sericea J.M.Wood & M. Evans			AF155243 / AF155280	-		Francisco Ortega & al., 2001
Gonospermum fruticosum Less.	Tenerife, Teno, Punta de Teno, Casa Blanca, 23.03.1999, Oberprieler 9877 (Herb. Oberprieler).			-	subm.	-
Helenium autumnale L.	-		AY688571S1	subm.		Simurda & al., 2005
Helichrysum lanceolatum Kirk	-		EU007682	-		Smissen & Breitwieser, 2008
Helichrysum lanceolatum Kirk				EF187698		Ford & al., 2007
Heliocauta atlantica (Litard. & Maire) Humphries	Ma., Toubkal, 3850 m, 23.08.1992, Kreisch 920589 (Herb. Kreisch).		AJ748782	subm.	subm.	Oberprieler, 2004a
Hippolytia dolichophylla (Kitam.) Bremer & Humphries			AJ748784	-	-	Oberprieler, 2004a
Leucanthemum vulgare (Vaill.) Lam. subsp. *pujiulae* Sennen	Frankreich: Dept. Pyrenees orientales, Haut Vallespir, La Clagere - S. Saveur im oberen Tal des Rio Tech westlich von Prats de Mollo la Preste, Silikat, ca. 850m, 24.8.1986, leg. R. Vogt 5053 & Ch. Prem (Herb. Vogt).		AJ3296398 / AJ3296433	subm.	subm.	Oberprieler & Vogt, 2000

Appendix A1: Continued.

Taxon	Accession	DIVA	EMBL ITS	EMBL psbA	EMBL trnC	Reference
Leucocyclus formosus Boiss.	Tu, Taunus, Bulgar Dagh, 6000', 09.07.1853, Kotschy n° 65 (M).		AJ864578	subm.	subm.	Oberprieler, 2004b
Lugoa revoluta (Link.) DC.	Tenerife, Anaga, Benijo, 26.03.1999, Oberprieler 9919 (Herb. Oberprieler).			subm.	subm.	-
Lugoa revoluta (Link.) DC.	-		AF155252 / AJ296412			Francisco Ortega & al., 2001
Matricaria discoidea DC.	Ge, Jena, Botanischer Garten, Oberprieler 9762 (cho, B).		AJ296447	subm.	subm.	Oberprieler & Vogt, 2000
Nananthea perpusilla DC.	Sardinien, Sulcis, Bucht NW Portoscuso, 0-20 m, 9.-20.4.1966, Merxmüller 21023 & Oberwinkler (M).	h	AJ864579	subm.		Oberprieler, 2004b
Pentzia dentata Kuntze	-		AY127681			Watson & al., 2002
Plagius maghrebinus Greuter & Vogt	Italia: Sardinia, surroundings of Sassari, Ittiri, near Bacino Cuga, 19.8.1996, leg. L. Zedda s.n. (Herb. Vogt).		AJ296438 /	subm.	subm.	Oberprieler & Vogt, 2000
Prolongoa hispanica Lopéz González & Jarvis	-		L77776			Francisco Ortega & al., 1997
Prolongoa hispanica Lopéz González & Jarvis	Spain, Málaga, Tolox - Sierra de las Nieves, 26.5.1992, Vogt 9233 (Herb. Vogt).			subm.	subm.	-
Senecio glaberrimus DC.	-		EF538338	EF538081		Pelser & al., 2007
Solidago gigantea Ait.	-		DQ005980	DQ006153	-	Kress & al., 2005
Soliva sessilis Ruiz Lopez & Pavon	-		AM774471	Himmelreich, unpubl. data	Himmelreich, unpubl. data	Himmelreich & al., 2008
Tagetes patula L.	-		DQ662121			Serrato-Cruz & al., unpublished
Tanacetum coccineum (Willd.) Grierson	Armenia, 12.06.2002, Oberprieler 10045 (cho).		AF155263	subm.	subm.	Francisco Ortega & al., 2001
Tanacetum vulgare L.	-	afhij	AJ864590	subm.	-	Oberprieler, 2004b
Tripleurospermum caucasicum (Willd.) Hayek	Armenia, 30.06.2002, Oberprieler 10192 (cho).			Himmelreich, unpubl. data	Himmelreich, unpubl. data	
Ursinia anthemoides (L.) Poir.	-		AM774473	Himmelreich, unpubl. data	Himmelreich, unpubl. data	Himmelreich & al., 2008

Appendix A2: Summary of the classification in sections and series for the taxa analysed in this thesis.

Taxon	Summary	Boissier - 1875 Sections	Boissier - 1875 Series	Eig - 1938	Fedorov - 1961 Sections	Fedorov - 1961 Series	Yavin - 1970-1972 Sections	Yavin - 1970-1972 Series	Grierson, Yavin - 1975 Sections	Fernandes - 1976 Subgen.	Fernandes - 1976 Sections	Iranshahr - 1986 Sections	Oberprieler - 1998 Subgen.	Oberprieler - 1998 Sections	Oberprieler - 1998 Series	Others
Anthemis aaronsohnii	Rascheyana Yavin			The group of A. rascheyana			Rascheyana Yavin									
Anthemis abrotanifolia	Hiorthia (DC.) Fern.									Anthemis	Hiorthia (DC.) Fern.					
Anthemis abylaea	Hiorthia (DC.) Fern.												Anthemis	Hiorthia (DC.) Fern.		
Anthemis aciphylla	Hiorthia (DC.) Fern.	Euanthemis	Perennes						Anthemis							
Anthemis adonidifolia	Maruta (Cass.) Griseb.	Maruta					Maruta (Cass.) Griseb.	Cotulae Fed.	Maruta (Cass.) Griseb.	Anthemis	Hiorthia (DC.) Fern.					
Anthemis aetnensis	Hiorthia (DC.) Fern.									Anthemis	Hiorthia (DC.) Fern.					
Anthemis alpestris	Hiorthia (DC.) Fern.									Anthemis	Hiorthia (DC.) Fern.					
Anthemis ammanthus	ser. Ammanthus									I.S.						
Anthemis ammophila	Anthemis	Euanthemis	Annuae vel Biennes				Anthemis	Ammanthus (Boiss.) Yavin	Anthemis	Anthemis	Anthemis					
Anthemis anthemiformis	Hiorthia (DC.) Fern.	Hiorthia (DC.) Fern.					Anthemis	Hyalinae Yavin	Anthemis	Anthemis	Anthemis					
Anthemis arnicola	Anthemis	Euanthemis	Annuae vel Biennes				Anthemis	Arnicolae Yavin	Anthemis	Anthemis	Anthemis					
Anthemis arvensis	Anthemis	Euanthemis	Annuae vel Biennes		Anthemis	Arvenses Fed.			Anthemis	Anthemis	Anthemis		Anthemis	Anthemis	Anthemis	
Anthemis atropaliana	Anthemis									Anthemis	Anthemis	Anthemis				
Anthemis auriculata	Anthemis	Euanthemis	Annuae vel Biennes				Anthemis	Tomentulosae Yavin	Anthemis	Anthemis	Anthemis					
Anthemis austroiranica	Rascheyana Yavin	Hiorthia (DC.) Fern.		The group of A. cotula			Rascheyana Yavin			Anthemis	Anthemis	Anthemis				
Anthemis bornmuelleri	Maruta (Cass.) Griseb.						Maruta (Cass.) Griseb.	Cotulae Fed.								
Anthemis bourgaei	Maruta (Cass.) Griseb.	Maruta					Anthemis	Bourgaeianae Yavin		Anthemis	Maruta (Cass.) Griseb.					
Anthemis boveana	Anthemis						Anthemis	Chrysanthae Yavin					Anthemis	Anthemis	Chrysanthae Yavin	
Anthemis brachystephana	Rascheyana Yavin						Rascheyana Yavin					Anthemis				

Appendix A2: Continued.

Taxon	Summary	Boissier - 1875 Sections	Boissier - 1875 Series	Eig - 1938	Fedorov - 1951 Sections	Fedorov - 1951 Series	Yavin - 1970-1972 Sections	Yavin - 1970-1972 Series	Grierson, Yavin - 1975 Sections	Fernandes - 1976 Subgen.	Fernandes - 1976 Sections	Iranshahr - 1986 Sections	Oberprieler - 1998 Subgen.	Oberprieler - 1998 Sections	Oberprieler - 1998 Series	Others
Anthemis breviradiata	Anthemis			The group of A. weilsiana			Anthemis	Melampodinae Yavin								
Anthemis bushehrica	Anthemis											Anthemis				
Anthemis calcarea	Hiorthia (DC.) Fern.	Euanthemis	Annuae vel Biennes		Ruralia Fed	Fruticuloae Fed			Anthemis			Anthemis				
Anthemis candidissima	Anthemis	Euanthemis	Annuae vel Biennes		Anthemis	Candidissimae Fed			Anthemis							
Anthemis chia	Cha Yavin	Euanthemis	Annuae vel Biennes	The group of A. chia			Cha Yavin		Anthemis	Anthemis	Cha Yav.		Anthemis	Anthemis	Secundiramese Yavin	
Anthemis confusa	Anthemis						Anthemis	Cornucopiae Yavin								
Anthemis cornucopiae	Anthemis	Euanthemis	Annuae vel Biennes	The group of A. cornucopiae			Anthemis		Anthemis							
Anthemis cotula	Marula (Cass.) Griseb.	Marula		The group of A. cotula	Marula (Cass.) Griseb.	Cotulae Fed	Marula (Cass.) Griseb.		Marula (Cass.) Griseb.	Anthemis	Marula (Cass.) Griseb.	Marula (Cass.) Griseb.	Anthemis	Marula (Cass.) Griseb.		
Anthemis cretica	Hiorthia (DC.) Fern.	Euanthemis	Perennes						Anthemis	Anthemis	Hiorthia (DC.) Fern.	Ruralia Fed	Anthemis	Hiorthia (DC.) Fern.		
Anthemis cretica subsp. *cretica*	Hiorthia (DC.) Fern.								Anthemis	Anthemis	Hiorthia (DC.) Fern.	Ruralia Fed				
Anthemis cretica subsp. *anatolica*	Hiorthia (DC.) Fern.	Euanthemis	Perennes						Anthemis	Anthemis	Hiorthia (DC.) Fern.	Ruralia Fed				
Anthemis cretica subsp. *argaea*	Hiorthia (DC.) Fern.	Euanthemis	Perennes						Anthemis	Anthemis	Hiorthia (DC.) Fern.	Ruralia Fed				
Anthemis cretica subsp. *carpatica*	Hiorthia (DC.) Fern.								Anthemis	Anthemis	Hiorthia (DC.) Fern.	Ruralia Fed				
Anthemis cretica subsp. *gerardiana*	Hiorthia (DC.) Fern.	Euanthemis	Perennes		Ruralia Fed	Anabicae Fed			Anthemis	Anthemis	Hiorthia (DC.) Fern.	Ruralia Fed				
Anthemis cretica subsp. *iblithorpei*	Hiorthia (DC.) Fern.								Anthemis	Anthemis	Hiorthia (DC.) Fern.	Ruralia Fed				
Anthemis cretica subsp. *spruneri*	Hiorthia (DC.) Fern.				Ruralia Fed	Saporlanae Fed			Anthemis	Anthemis	Hiorthia (DC.) Fern.	Ruralia Fed				
Anthemis cuneata	Hiorthia (DC.) Fern.								Anthemis	Anthemis	Hiorthia (DC.) Fern.					
Anthemis cupaniana	Anthemis						Anthemis	Secundiramese Yavin	Anthemis				Anthemis	Anthemis	Secundiramese Yavin	
Anthemis cyrenaica	Anthemis						Anthemis	Hauschnechtianae Yavin								
Anthemis davisii	Anthemis			The group of A. weilsiana			Anthemis	Yavin								
Anthemis desert-syriaca	Anthemis						Anthemis	Melampodinae Yavin								
Anthemis edumea	Anthemis			The group of A. weilsiana			Anthemis	Melampodinae Yavin								

Appendix A2: Continued.

Taxon	Summary	Boissier - 1875 Sections	Boissier - 1875 Series	Eig - 1938	Fedorov - 1961 Sections	Fedorov - 1961 Series	Yavin - 1970-1972 Sections	Yavin - 1970-1972 Series	Grierson, Yavin - 1975 Sections	Fernandes - 1976 Subgen.	Fernandes - 1976 Sections	Iranshahr - 1986 Sections	Oberprieler - 1996 Subgen.	Oberprieler - 1996 Sections	Oberprieler - 1996 Series	Others
Anthemis filicaulis	ser. Ammanthus						Anthemis	Ammanthus (Boiss.) Yavin		Ammanthus (Boiss. & Heldr.) R.Fern.						
Anthemis fimbriata	Anthemis	Euanthemis	Annuae vel Biennes				Anthemis	Hyalinae Yavin	Anthemis							
Anthemis freitagi	Anthemis	Euanthemis	Perennes			Fruticulosae Fed	Anthemis	Roseae Yavin				Anthemis				
Anthemis fruticulosa	Hiorthia (DC.) Fern.	Euanthemis	Annuae vel Biennes	The group of A. montana Eig	Rumata Fed	Fruticulosae Fed										
Anthemis fumariifolia	Anthemis	Euanthemis	Annuae vel Biennes				Anthemis					Anthemis				
Anthemis gayana	Rascheyana Yavin						Rascheyana Yavin					Anthemis				
Anthemis ghilanensis	Anthemis						Rascheyana Yavin						Anthemis	Anthemis	Chrysanthae Yavin	
Anthemis glauca	Rascheyana Yavin											Anthemis				
Anthemis glaberrima	ser. Ammanthus						Anthemis	Ammanthus (Boiss.) Yavin		LS			Anthemis	Anthemis	Secundiramae Yavin	
Anthemis glaucosa	Anthemis						Anthemis	Secundiramea Yavin				Anthemis				
Anthemis hamrinensis	Anthemis						Anthemis									
Anthemis haroldi-mazzettii	Anthemis			The group of A. hyalina			Anthemis	Melampodinae Yavin				Anthemis				
Anthemis haussknechtii	Anthemis	Euanthemis	Annuae vel Biennes	The group of A. haussknechtii			Anthemis	Haussknechtianae Yavin	Anthemis							
Anthemis halonica	Anthemis	Euanthemis	Annuae vel Biennes	The group of A. cornuciopiae			Anthemis	Cornuciopae Yavin								
Anthemis hebronica	Anthemis				Anthemis	Candidissimae Fed										
Anthemis tenella	Anthemis			The group of A. wettsteiniana			Rascheyana Yavin					Anthemis				
Anthemis humailalepis	Rascheyana Yavin	Euanthemis	Annuae vel Biennes	The group of A. hyalina			Anthemis	Hyalinae Yavin	Anthemis	Anthemis	Hiorthia (DC.) Fern.					
Anthemis hyalina	Hiorthia (DC.) Fern.						Anthemis	Hyalinae Yavin				Anthemis				
Anthemis hydruntina	Anthemis	Euanthemis	Annuae vel Biennes	The group of A. hyalina A. melampodina			Anthemis	Induratae Yavin								
Anthemis indurata	Hiorthia (DC.) Fern.						Anthemis			Anthemis	Hiorthia (DC.) Fern.	Anthemis				
Anthemis icmelia	Anthemis															
Anthemis karsitanica																

Appendix A2: Continued.

Taxon	Summary	Boissier - 1875		Eig - 1938	Fedorov - 1961		Yavin - 1970-1972		Grierson, Yavin - 1975	Fernandes - 1976		Iranshahr - 1986	Oberprieler - 1998			Others
		Sections	Series		Sections	Series	Sections	Series	Sections	Subgen.	Sections	Sections	Subgen.	Sections	Series	
Anthemis kotschyana	Hortha (DC) Fern.	Euanthemis	Perennes						Anthemis			Rumata Fed				(Franzen 1986)
Anthemis laconica	Anthemis	Euanthemis														
Anthemis leptophylla	Anthemis			The group of A. hyalina			Anthemis	Hyalinae Yavin				Anthemis				
Anthemis leucanthemifolia	Anthemis	Euanthemis	Annuae vel Biennis	The group of A. hyalina			Anthemis	Induratae Yavin				Anthemis				
Anthemis lorestanica	Anthemis	Euanthemis	Annuae vel Biennis													
Anthemis macedonica	Hortha (DC) Fern.				Anthemis	Anthemis				Anthemis	Anthemis					
Anthemis macrotis	Manula (Cass) Griseb															Manula (Cass.) Griseb. (Oberprieler & Vogt 2006)
Anthemis maritima	Hortha (DC) Fern.	Euanthemis	Perennes		Anthemis	Hortha (DC) Fern.							Anthemis	Hortha (DC) Fern.	Chrysanthae Yavin	
Anthemis maroccana	Anthemis												Anthemis	Anthemis		
Anthemis marschalliana	Hortha (DC) Fern.	Euanthemis	Perennes						Anthemis				Anthemis	Anthemis	Bourgaeanianae Yavin	
Anthemis mauritiana	Anthemis	Euanthemis	Annuae vel Biennis	The group of A. melampodina			Anthemis	Bourgaeanianae Yavin								
Anthemis melampodina	Anthemis	Euanthemis	Annuae vel Biennis				Anthemis	Melampodinae Yavin								
Anthemis melanolome	Anthemis	Euanthemis	Annuae vel Biennis				Anthemis	Hauszknechtianae Yavin	Anthemis	Anthemis	Hortha (DC) Fern.					
Anthemis melitonica	Hortha (DC) Fern.															
Anthemis microcephala	Manula (Cass) Griseb			The group of A. cotula	Manula (Cass) Griseb	Microcephalae Fed	Manula (Cass) Griseb	Microcephalae Fed				Manula (Cass.) Griseb.				
Anthemis microlepis	Anthemis			The group of A. cornuspcae			Anthemis	Hyalinae Yavin				Anthemis				
Anthemis microsperma	Anthemis	Euanthemis	Annuae vel Biennis				Anthemis	Induratae Yavin								
Anthemis mirheydari	Anthemis						Anthemis	Chrysanthae Yavin			Anthemis	Anthemis				
Anthemis moghanica	Anthemis											Anthemis				
Anthemis muricata	Anthemis															

Appendix A2: Continued.

Taxon	Summary	Boissier - 1875 Sections	Boissier - 1875 Series	Eig - 1938	Fedorov - 1961 Sections	Fedorov - 1961 Series	Yavin - 1970-1972 Sections	Yavin - 1970-1972 Series	Grierson, Yavin - 1975 Sections	Fernandes - 1976 Subgen.	Fernandes - 1976 Sections	Iranshahr - 1986 Sections	Oberprieler - 1998 Subgen.	Oberprieler - 1998 Sections	Oberprieler - 1998 Series	Others
Anthemis nabataea	Anthemis			The group of A. cornucopiae			Anthemis	Melampodinae Yavin								
Anthemis odontostephana	Odontostephana Eig	Maruta	Perennes	Sec. Odontostephana Eig	Maruta (Cass.) Griseb.	Odontostephanae Fed.	Odontostephana Eig					Odontostephana Eig		Odontostephana Eig		Anthemis (Meikle 1963)
Anthemis orientalis	Hortha (DC) Fern.	Euanthemis	Perennes						Anthemis	Anthemis	Hortha (DC) Fern.					
Anthemis parnesa	Anthemis	Euanthemis	Annuae vel Biennes	The group of A. montana			Anthemis	Tomenticae Yavin	Anthemis	Anthemis	Anthemis	Rumata Fed.				
Anthemis pauciloba	Hortha (DC) Fern.									Anthemis	Hortha (DC) Fern.		Anthemis	Hortha (DC.) Fern.		
Anthemis pedunculata	Hortha (DC) Fern.	Euanthemis	Annuae vel Biennes							Anthemis	Anthemis					
Anthemis peregrina	Anthemis	Euanthemis	Annuae vel Biennes									Anthemis				
Anthemis persica	Anthemis	Euanthemis	Annuae vel Biennes				Anthemis	Hyalinae Yavin								
Anthemis plutonia	Hortha (DC) Fern.															
Anthemis pseudocotula	Maruta (Cass.) Griseb.	Maruta	Annuae vel Biennes	The group of A. cotula			Maruta (Cass.) Griseb.	Cotulae Fed.	Maruta (Cass.) Griseb.	Anthemis	Maruta (Cass.) Griseb.	Maruta (Cass.) Griseb.	Anthemis	Maruta (Cass.) Griseb.		
Anthemis punctata	Hortha (DC) Fern.									Anthemis	Hortha (DC) Fern.		Anthemis	Hortha (DC.) Fern.		
Anthemis rascheyana	Rascheyana Yavin	Euanthemis	Annuae vel Biennes				Rascheyana Yavin									
Anthemis retusa	Maruta (Cass.) Griseb.	Maruta	Annuae vel Biennes	The group of A. rascheyana			Maruta (Cass.) Griseb.	Cotulae Fed.								
Anthemis rhodocentra	Anthemis					Arvensis Fed.						Anthemis				
Anthemis rigida	Anthemis	Euanthemis	Annuae vel Biennes				Anthemis	Securidurineae Yavin	Anthemis	Anthemis	Anthemis					
Anthemis rumelica	Hortha (DC) Fern.	Euanthemis	Annuae vel Biennes							Anthemis	Hortha (DC) Fern.					
Anthemis ruthenica	Anthemis	Euanthemis	Annuae vel Biennes				Anthemis			Anthemis	Anthemis	Anthemis				
Anthemis scariosa	Anthemis	Euanthemis	Annuae vel Biennes	The group of A. scariosa	Anthemis		Anthemis	Scariosae Yavin	Maruta/Anthemis	Anthemis	Anthemis	Anthemis				
Anthemis schizostephana	Rascheyana Yavin	Euanthemis	Annuae vel Biennes				Rascheyana Yavin									
Anthemis scopulorum	Anthemis						Anthemis	Tomenticae Yavin								

Appendix A2: Continued.

Taxon	Summary	Boissier - 1875 Sections	Boissier - 1875 Series	Eig - 1938	Fedorov - 1961 Sections	Fedorov - 1961 Series	Yavin - 1970-1972 Sections	Yavin - 1970-1972 Series	Grierson, Yavin - 1975 Sections	Fernandes - 1976 Subgen	Fernandes - 1976 Sections	Iranshahr - 1986 Sections	Oberprieler - 1998 Subgen	Oberprieler - 1998 Sections	Oberprieler - 1998 Series	Others
Anthemis scrobicularis	Anthemis						Anthemis	Melampodinae Yavin								
Anthemis secundiramea	Anthemis	Euanthemis	Annuae vel Biennes				Anthemis	Secundirameae Yavin		Anthemis	Anthemis		Anthemis	Anthemis	Secundirameae Yavin	
Anthemis secundiramea subsp. *urvilleana*	Anthemis	Euanthemis	Annuae vel Biennes		Runalia Fed.	Anatolicae Fed.	Anthemis	Secundirameae Yavin		Anthemis	Anthemis		Anthemis	Anthemis	Secundirameae Yavin	
Anthemis sintenisii	Hiorthia (DC.) Fern.								Anthemis	Anthemis	Hiorthia (DC.) Fern.	Anthemis				
Anthemis sterilis	Anthemis	Euanthemis	Perennes				Anthemis	Anrinacleae Yavin					Anthemis	Anthemis	Chrysanthae Yavin	
Anthemis susiana	Anthemis	Euanthemis	Annuae vel Biennes													
Anthemis tenuisecta	Maruta (Cass.) Griseb.															
Anthemis tigrensis	ser Ammanthus						Anthemis	Ammanthus (Boiss.) Yavin		Ammanthus (Boiss. & Heldr.) R.Fern.	Anthemis			Anthemis		
Anthemis tomentella	Anthemis	Euanthemis	Perennes				Anthemis	Tomentosae Yavin	Anthemis	Anthemis	Anthemis					
Anthemis tomentosa	Hiorthia (DC.) Fern.	Euanthemis	Perennes				Anthemis	Anrinacleae Yavin	Anthemis	Anthemis	Hiorthia (DC.) Fern.					
Anthemis tricolr	Anthemis	Euanthemis	Perennes	The group of A. cornucopiae			Anthemis	Anrinacleae Yavin	Anthemis							
Anthemis tricornis	Maruta (Cass.) Griseb.	Maruta		The group of A. cotula			Maruta (Cass.) Griseb.	Cotulae Fed.								
Anthemis tripolitana	Hiorthia (DC.) Fern.	Euanthemis	Perennes	Sec. Odontostephana Eig	Runalia Fed.	Fruticuloae Fed.	Odontostephana Eig			Anthemis	Hiorthia (DC.) Fern.					
Anthemis trotzkiana	Odontostephana Eig	Maruta			Maruta (Cass.) Griseb.	Odontostephana Fed.	Odontostephana Eig									
Anthemis tubicina	Anthemis						Anthemis	Tomentosae Yavin								
Anthemis ubensis	Anthemis						Anthemis	Melampodinae Yavin		Anthemis	Anthemis	Anthemis				
Anthemis werneri	Maruta (Cass.) Griseb.			The group of A. wettsteiniana									Anthemis	Anthemis	Secundirameae Yavin	
Anthemis wettsteiniana	Anthemis															
Anthemis yemensis	Maruta (Cass.) Griseb.															
Anthemis zaianica	Anthemis												Anthemis	Anthemis	Bourganianae Yavin	Anthemis (Prodisch) 1982
Anthemis zoharyana	Anthemis			The group of A. melampodina			Anthemis	Melampodinae Yavin								

Appendix A2: Continued.

Taxon	Summary	Boissier - 1875 Sections	Boissier - 1875 Series	Eig. - 1938	Fedorov - 1961 Sections	Fedorov - 1961 Series	Yavin - 1970-1972 Sections	Yavin - 1970-1972 Series	Grierson, Yavin - 1975 Sections	Fernandes - 1976 Subgen.	Fernandes - 1976 Sections	Iranshahr - 1986 Sections	Oberprieler - 1998 Subgen.	Oberprieler - 1998 Sections	Oberprieler - 1998 Series	Others
Cota altissima	Cota	Cota Boiss.		Sec. Cota Boiss	Cota Boiss.	Altissimae Fed.			Cota Boiss.	Cota	Cota	Cota				
Cota amblyolepis	Cota			Sec. Cota Boiss												
Cota austriaca	Cota	Cota Boiss.		Sec. Cota Boiss	Cota Boiss.	Altissimae Fed.			Cota Boiss.	Cota	Cota	Cota	Cota (J. Gay) Rouy	Cota (J. Gay) Rchb. F.		
Cota brevicuspis	Cota			Sec. Cota Boiss	Cota Boiss.	Altissimae Fed.						Cota				
Cota coelopoda	Cota	Cota Boiss.		Sec. Cota Boiss	Cota Boiss.	Altissimae Fed.			Cota Boiss.	Cota	Cota	Cota				
Cota dalmatica	Cota				Cota Boiss.				Cota Boiss.	Cota	Cota					
Cota dipsacea	Cota				Cota Boiss.	Rigescentes Fed.			Cota Boiss.							
Cota dubia	Anthemaria Dum.				Cota Boiss.	Rigescentes Fed.				Cota	Anthemaria Dum.					
Cota dumetorum	Anthemaria Dum.				Cota Boiss.	Rigescentes Fed.				Cota	Anthemaria Dum.					
Cota fulvida	Anthemaria Dum.								Cota Boiss.							
Cota halophila	Cota	Cota Boiss.		Sec. Cota Boiss	Cota Boiss.	Rigescentes Fed.	Anthemis	Halophilae Yavin	Cota Boiss.							
Cota linczevskyi	Anthemaria Dum.				Cota Boiss.	Rigescentes Fed.										
Cota macrantha	Anthemaria Dum.									Cota	Anthemaria Dum.					
Cota macroglossa	Anthemaria Dum.				Cota Boiss.	Macroglossae Fed.						Cota				
Cota mazandaranica	Cota				Cota Boiss.	Macroglossae Fed.										
Cota melanoloma	Anthemaria Dum.				Cota Boiss.	Tinctoriae Fed.			Cota Boiss.	Cota	Anthemaria Dum.					
Cota monantha	Anthemaria Dum.															
Cota oxylepis	Anthemaria Dum.	Cota Boiss.							Cota Boiss.	Cota	Cota	Cota				
Cota palaestina	Cota	Cota Boiss.							Cota Boiss.							
Cota pestalozzae	Cota	Cota Boiss.							Cota Boiss.							
Cota rigescens	Anthemaria Dum.	Cota Boiss.			Cota Boiss.	Rigescentes Fed.										
Cota seguramica	Anthemaria Dum.				Cota Boiss.	Tinctoriae Fed.										

Appendix A2: Continued.

Taxon	Summary	Boissier - 1875		Eig - 1938	Fedorov - 1961		Yavin - 1970-1972		Grierson, Yavin - 1975	Fernandes - 1976		Transhahr - 1986	Oberprieler - 1998			Others
		Sections	Series		Sections	Series	Sections	Series	Sections	Subgen.	Sections	Sections	Subgen.	Sections	Series	
Cota samuelssonii	Cota															Cota (Rech.f 1963)
Cota segetalis	Cota									Cota		Cota				
Cota tapyichensis	Anthemaria Dum.	Cota Boiss.		Sec. Cota Boiss	Cota Boiss	Pigosantes Fed			Cota Boiss.	Cota	Anthemaria Dum.	Cota				
Cota tinctoria	Anthemaria Dum.			Sec. Cota Boiss	Cota Boiss	Tinctoriae Fed			Cota Boiss.	Cota	Anthemaria Dum.	Cota				
Cota triumfettii	Anthemaria Dum.								Cota Boiss.			Cota				
Cota weidmanniana	Cota	Cota Boiss.			Cota Boiss	Weidmanniniae Fed			Cota Boiss.			Cota				

Appendix A3: Summary of the characters states obtained for all 46 taxa from the micromorphological analyses. Character names and character specifications are given in Table 2.1. (Modified from Oppolzer, 2007).

Character name abbreviation	M1	M2	M3	M4	M5	M6	M7	M8	M9	M10	M11	M12	M13	M14	M15	M16	M17	M18	M19	M20	M21	M22	M23	M24	M25
A. adonidifolia	2	0	2	1	2.5	1.10	1	0	1	1	2	0	1	4	0.40	3	7.0	20.0	1	1	0	1	2	1	1
A. ammanthus	2	1	1	2	2.5	1.00	0	0	2	2	1	2	1	2	0.29	3	7.3	18.0	1	1	1	1	2	1	1
A. arvensis	2	0	3	3	2.5	0.66	0	0	1	2	2	2	1	2	0.30	2	7.0	32.0	0	1	0	0	2	1	1
A. atropatana	2	0	2	1	3.0	1.00	0	0	1	2	2	2	1	2	0.33	3	7.0	18.3	?	1	0	1	2	1	1
A. bourgaei	3	1	2	2	1.5	0.75	0	0	2	2	1	?	1	1	0.26	2	6.33	28.0	0	?	1	?	2	1	1
A. calcarea	2	1	1	2	3.0	0.75	0	?	1	2	1	2	0	1	0.00	4	8.3	36.7	1	2	0	0	2	1	0
A. candidissima	2	0	3	1	2.5	0.75	0	0	2	2	2	2	1	4	0.30	1	8.33	28.0	?	?	?	?	?	?	?
A. chia	2	1	3	?	3.0	1.00	0	0	1	2	2	1	1	4	0.37	1	6.3	19.0	1	1	1	0	2	1	1
A. cotula	2	0	3	1	2.0	0.75	0	0	1	2	2	2	1	2	0.26	2	6.0	18.3	1	1	0	0	2	1	1
A. cuneata	3	0	1	1	3.0	0.90	0	1	2	2	2	2	0	1	0.00	2	7.0	23.0	0	1	0	0	2	1	1
A. liticaulis	2	1	1	1	2.5	1.00	0	?	1	2	1	2	1	2	0.32	4	5.7	22.3	1	1	1	1	2	1	1
A. gilanica	2	0	2	1	2.5	1.00	0	0	1	2	2	2	1	2	0.31	1	7.0	21.7	1	1	1	0	2	1	1
A. glaberrima	2	1	0	2	0.0	0.00	0	?	1	2	1	?	1	2	0.34	4	5.7	6.7	?	1	?	1	2	1	1
A. handel-mazzettii	2	0	3	3	1.5	1.00	0	0	1	1	2	2	1	2	0.47	3	7.0	23.7	0	1	1	0	2	1	1
A. homalolepis	2	0	2	1	2.0	0.90	0	0	1	2	2	2	1	1	0.34	4	5.3	22.0	1	1	1	0	2	1	1
A. hyalina	3	0	2	2	3.0	1.00	0	0	1	2	2	0	1	4	0.41	1	9.33	20.7	1	1	1	0	?	1	1
A. leucanthemifolia	2	1	3	1	2.5	0.66	0	0	1	1	2	2	1	1	0.30	2	7.0	18.3	1	1	0	1	2	1	1
A. macedonica	2	0	?	1	2.5	0.75	0	0	2	2	2	?	1	2	0.29	4	7.67	31.7	1	?	?	?	2	1	1
A. maritima	3	1	0	1	3.5	0.90	0	0	1	2	2	2	1	1	0.26	3	10.3	35.3	1	1	0	0	2	1	1
A. marschalliana	2	1	1	2	3.5	0.75	1	1	1	2	1	2	1	2	0.00	2	8.3	34.3	1	1	0	0	2	1	1
A. microcephala	2	?	1	1	2.0	0.66	0	0	1	2	2	0	0	3	0.00	2	5.7	23.7	0	1	1	0	2	1	1
A. muricata	0	1	2	1	3.0	1.00	0	0	2	2	1	?	1	4	0.36	1	9.67	26.3	1	?	1	1	2	1	1
A. odontostephana	3	1	2	2	3.0	1.00	0	0	2	2	1	0	0	3	0.00	2	6.33	17.3	1	1	1	0	2	1	1
A. pedunculata	3	0	0	1	2.0	0.75	0	0	1	2	1	2	1	2	0.32	2	8.33	34.7	1	?	?	?	2	1	?
A. persica	2	0	2	2	2.5	1.00	0	0	1	2	2	?	1	1	0.26	2	7.33	25.0	1	?	1	0	2	?	1
A. pseudocotula	2	0	3	3	2.0	1.00	0	0	1	2	2	2	1	2	0.30	2	7.00	31.3	1	?	0	0	2	1	1
A. rascheyana	2	1	2	1	2.0	1.00	0	0	1	1	2	2	1	1	0.25	4	7.3	26.3	1	1	0	0	2	1	1
A. rigida	2	1	1	3	3.0	1.00	0	1	1	2	2	0	1	1	0.29	4	7.0	26.7	1	1	0	0	2	1	1
A. rosea	2	0	3	2	3.0	0.66	0	0	1	1	2	2	1	4	0.34	2	7.3	25.3	1	2	0	0	2	1	1
A. secundiramea	0	1	0	3	2.5	1.00	0	0	1	2	2	2	1	2	0.34	2	6.33	23.3	1	?	?	0	2	1	?
A. susiana	2	1	3	1	2.5	1.00	0	0	2	2	1	1	1	1	0.34	3	7.0	21.7	?	1	1	1	2	1	1
A. tigrensis	3	1	2	0	2.0	0.66	?	?	1	2	1	?	1	2	0.34	4	7.33	30.3	1	?	0	0	2	1	?
A. tomentella	2	1	1	3	3.5	1.00	?	0	1	2	2	?	1	2	0.32	2	7.7	27.7	1	1	1	1	2	1	1
A. tricolor	2	0	2	1	4.0	0.75	0	0	1	2	1	0	1	1	0.29	3	7.33	29.7	1	?	?	?	?	?	?
A. trotzkiana	1	1	2	2	3.0	0.90	0	1	1	2	1	2	0	1	0.00	2	9.0	30.0	1	1	0	0	2	1	1
A. tubicina	3	0	2	3	2.0	1.00	0	0	2	2	1	0	0	3	0.00	1	7.7	17.3	0	1	1	0	2	1	1
C. altissima	1	0	3	1	3.0	1.00	0	0	1	2	2	1	1	1	0.23	3	7.7	21.0	1	2	0	0	3	2	0
C. dumetorum	2	0	3	1	2.5	1.00	0	0	2	2	1	1	1	1	0.35	3	7.7	26.0	0	2	0	0	1	2	0
C. halophila	2	0	3	3	1.5	0.66	0	0	1	2	2	2	1	2	0.46	2	4.7	16.7	1	1	0	0	2	1	1
C. palaestina	1	1	3	3	3.0	1.00	0	0	1	2	2	0	1	1	0.26	2	10.00	17.0	1	?	?	?	1	2	?
C. samuelssonii	3	0	1	1	1.5	0.66	0	0	1	2	2	1	1	1	0.37	2	6.3	16.7	?	1	0	0	1	1	1
C. segetalis	1	0	3	2	2.5	1.00	0	0	1	1	1	1	1	1	0.37	2	7.0	24.3	?	2	0	0	3	2	1
C. triumfettii	2	0	3	2	2.0	1.00	0	0	1	2	2	1	1	2	0.38	2	7.7	24.0	?	2	0	0	3	2	1
N. perpusilla	0	1	0	0	3.0	1.10	0	0	2	2	1	0	0	4	0.00	2	7.0	21.3	0	1	1	0	3	1	0
T. parviflorum	1	1	3	1	2.5	0.75	0	0	2	2	1	2	0	1	0.00	2	7.0	21.0	1	?	1	0	1	1	1
T. perforatum	0	1	3	1	2.0	0.66	0	1	2	2	1	0	0	1	0.00	2	6.0	22.3	1	?	0	0	1	1	1

Appendix A4: Proposed new combinations for six taxa belonging to *Cota*.

Cota brevicuspis (Bornm.) Lo Presti & Oberprieler, comb. nova ≡ *Anthemis brevicuspis* Bornm. in Beih. Bot. Centrbl. 32, 2: 397. 1914.

Cota linczevskyi (Fedor.) Lo Presti & Oberprieler, comb. nova ≡ *Anthemis linczevskyi* Fedor. in Fl. USSR 25: 868. 1961.

Cota macroglossa (Somm. & Lev.) Lo Presti & Oberprieler, comb. nova ≡ *Anthemis macroglossa* Somm. & Lev. in Nuov. Giorn. Bot. Ital. 2, 2: 85. 1895.

Cota mazandaranica (Iranshahr) Lo Presti & Oberprieler, comb. nova ≡ *Anthemis mazandaranica* Iranshahr in Pl. Syst. Evol. 139: 317. 1982.

Cota saguramica (Sosn.) Lo Presti & Oberprieler, comb. nova ≡ *Anthemis saguramica* Sosn. in Monit. Jard. Bot. Tiflis, 1926-27, n. s. Pt. 3-4, 150 (1927).

Cota talyschensis (Fedor.) Lo Presti & Oberprieler, comb. nova ≡ *Anthemis talyschensis* Fedor. in Fl. USSR 25: 869. 1961.

Next page: **Appendix A5**: Support for areas reconstruction based on a likelihood optimisation conducted on a sample of 1,000 trees obtained from the bayesian analysis and made ultrametric. The support is shown on the ultrametric ML tree based on nrDNA ITS data (see chapter 3). The support for each area is shown in brackets and is calculated as the product of topological uncertainty (measured as the posterior probability) and of area optimisation uncertainty (measured as the average of likelihood over 1,000 trees). Only support values ≥ 0.70 are shown. a) Lower part of the tree; b) (next page) upper part of the tree.

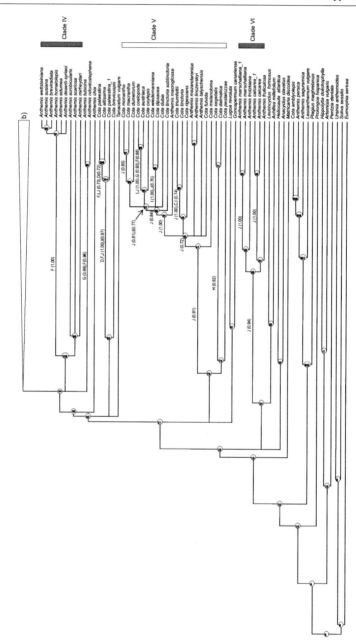

Appendix A5: Continued.

b)

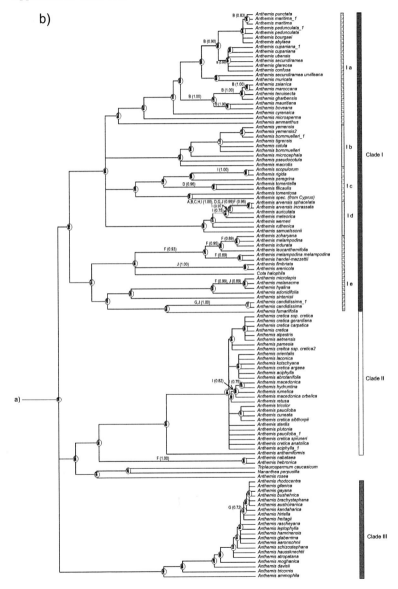

Appendix A6: Key to 147 of the 152 *Anthemis* species listed in Appendix A1. For the following taxa no detailed information was available: *A. debilifolia*, *A. fumarioides*, *A. laconica*, *A. handel-mazzettii*, *A. homalolepis*.

1.	Paleae oblong or oblanceolate, abruptly and shortly acute or acuminate; achenes ± angular or squarish in section, usually auriculate at apex, sometimes rounded; annuals and perennials	2
1a.	Paleae narrowly linear-lanceolate or subulate, 10-15 x longer than broad; achenes ± terete (cylindrical and circular in cross section), apex rounded, lobed or denticulate, not strongly auriculate	172
2.	Plants annual	3
2a.	Plants perennial	119
3.	Receptacle without scales	4
3a.	Receptacle with scales at least in the upper half	5
4.	Leaves fleshy, the lobes obtuse; achenes all similar, obconical-turbinate, distinctly ribbed, caducous	*A. ammanthus*
4a.	Leaves not fleshy, the lobes acute; outer achenes cylindrical, obscurely ribbed, persistent, the inner cylindric-obconical, distinctly ribbed, caducous	*A. filicaulis*
5.	Ligules yellow	6
5.	Ligules white, pinkish or absent	13
6.	Achenes smooth	7
6a.	Achenes tuberculate	9
7.	Leaves all basal	*A. gracilis*
7a.	Leaves throughout the whole stem	8
8.	Achenes with strongly elongate mucilage cells forming continuous slim ribbons on the ridges of ribs	*A. gharbensis*
8a.	Achenes with isodiametric or only slightly elongate mucilage cells that never form continuous slim ribbons on the ridges of ribs	*A. tenuisecta*
9.	Pales with tips formed by the protruding midribs	10
9a.	Midrib of pales never reaching the apex	11
10.	Limb of ray florets shorter than 7 mm; leaves usually densely tomentose; ultimate leaf segments broadly elliptical to nearly circular in outline	*A. chrysantha*
10a.	Limb of ray florets longer than 7 mm; leaves usually sparsely hairy; ultimate leaf segments elliptical	*A. boveana*
11.	Leaves petiolate	*A. maroccana*
11a.	Leaves sessile	12
12.	Achenes up to 1,7 mm long, heteromorphic	*A. tenuisecta*
12a.	Achenes 2-3 mm long, omomorphic	*A. scariosa*
13.	Ligules purplish-pink at least at base	14
13a.	Ligules white or absent	16
14.	Ligules evidently pink	15
14a.	Ligules white but purplish at base; disc florets pinkish	*A. freitagii*
15.	Stem glabrous	*A. glaberrima*
15a.	Stem pubescent	*A. rosea*

16.   Ligules absent or up to 2-3 white ligules per capitulum   17
16a.   Ligules white   23
17.   Receptacle without scales   *A. ammanthus*
17a.   Receptacle with scales at least in the upper half   18
18.   Receptacle without scales in the lower part; scales narrowly lanceolate
        to linear-subulate   *A. tomentella*
18a.   Receptacle with scales all over   19
19.   Outer and inner involucral bracts of the same form   20
19a.   Outer involucral bracts acute to triangular, inner ones obtuse or ovate   21
20.   Stems procumbent; achenes tuberculate; disc floret corolla 2-3 mm long
                                                                              A. persica
20a.   Stems rigid; achenes auricled; disc floret corolla 3-3,5 mm long   *A. rigida*
21.   Leaves pinnatipartite, 1-2 pinnatipartite to 1-2 pinnatisect; achenes with ribs
                                                                                       22
21a.   Leaves 2-3 pinnatipartite to 2-3 pinnatisect; stems erect; involucre strongly
        umbonate   *A. cyrenaica*
22.   Achenes of outer florets subcylindrical, coronate, ca. 1,7 x 0,8 mm;
        sometimes 1,3 white ligules pro capitulum   *A. kruegeriana*
22a.   Achenes of outer florets subprismatic ca. 1,5-2 x 1,25-1,5 mm, with a rather
        thick auricle up to 2/3 achene   *A. muricata*
23.   Outer   and/or   inner   achenes   evidently   tuberculate,   tuberous   or
        subtuberculate   24
23a.   Outer and/or inner achenes not tuberculate or with smooth ribs   48
24.   Outer and inner involucral bracts of the same form   25
24a.   Outer involucral bracts acute to triangular, inner ones obtuse or ovate   40
25.   Peduncles thickened at last in fruit   26
25a.   Peduncles never thickened   31
26.   Corolla tube of disc floret inflated   27
26a.   Corolla tube of disc floret not inflated   30
27.   Outer and/or inner achenes auriculated   28
27a.   All achenes without auricles   *A. nabatea*
28.   Auricles only in disc achenes; receptacle hemispherical   *A. mirehydari*
28a.   Auricles in all achenes; receptacle conical   29
29.   Achenes lenticular-punctate, subtuberculate; auricles crenate or denticulate
                                                                              A. fumariifolia
29a.   Achenes evidently tuberculate; auricles lacerate   *A. sintenisii*
30.   Leaves 3-pinnatisect; receptacle conical; ligules as long as disc
                                                                              A. haussknechtii
30a.   Leaves 1-2 pinnatisect; receptacle hemispherical; ligule as long as 3x disc
                                                                              A. mirehydari
31.   Corolla tube of disc floret inflated or slightly inflated   33
31a.   Corolla tube of disc floret not inflated   39
32.   Outer involucral bracts white- or hyaline-margined   33
32a.   Outer involucral bracts dark-margined, achenes with long auricles
                                                                              A. sintenisii
33.   Outer and/or inner achenes auriculated   34
33a.   Achenes with very short auricle, or without auricles   38

34.	Auricles only in disc achenes, as long as achene	*A. mirheydari*
34a.	Auricles in all achenes	35
35.	Auricles of outer achenes as long as achenes	*A. cornucopiae*
35a.	Auricles as long as 1/2 - 1/3 achenes, crenate, eroded or lacerate	36
36.	Outer achenes up to 2,5 mm long	37
36a.	Outer achenes 2,5 – 3,5 mm	*A. hebronica*
37.	Achenes 1 – 1,25 mm long; auricle crenate or eroded	*A. brachycarpa*
37a.	Achenes 1,5 – 2,5 mm long; auricle lacerate	*A. sintenisii*
38.	Disc florets glabrous, plant gray-pubescent	*A. scrobicularis*
38a.	Disc florets hairy, plant not pubescent	*A. maris-mortui*
39.	Capitula c. 15 mm diameter; receptacle hemispherical; disc achenes with auricles	*A. mirheydari*
39a.	Capitula c. 20-25 mm diameter; receptacle conical; all achenes without auricles	*A. scrobicularis*
40.	Auricles in all achenes or only in outer ones	41
40a.	Auricles absent or very short (up to 0,6 mm) and only in inner achenes	46
41.	Auricles as long as achenes	42
41a.	Auricles as long as 1/2 achene or shorter (at least in outer achenes)	43
42.	Auricles entire	*A. melampodina*
42a.	Auricles 3-5 lobate	*A. zoharyana*
43.	Stem procumbent; outer achenes not tuberculate	*A. glareosa*
43a.	Stem erect or ascending	44
44.	Basal tube of ray florets glabrous	45
44a.	Basal tube of ray florets villous	*A. davisii*
45.	Involucral bracts 4 mm or longer	*A. wettsteiniana*
45a.	Involucral bracts up to 4 mm	*A. edumea*
46.	Pales deciduous at maturity	47
46a.	Pales persistent at maturity; corona up to 0,1 mm; outer achenes not tuberculate	*A. confusa*
47.	Auricle absent; achenes up to 1 mm long	*A. bourgaei*
47a.	Corona or auricle up to 0,6 mm; achenes at least 1.5 mm	*A. stiparum*
48.	Outer and inner involucral bracts of the same form	49
48a.	Outer involucral bracts acute to triangular, inner ones obtuse or ovate	77
49.	Peduncles clavate at maturity	50
49a.	Peduncles not clavate at maturity	65
50.	Achenes with auricle or crown	51
50a.	Achenes without auricle at all	58
51.	Involucral bracts dark margined	*A. haussknechtii*
51a.	Involucral bracts hyaline or pale margined	52
52.	Tube of disc florets violaceus; disc achenes narrowly crowned	53
52a.	Tube of disc florets not violaceus	54
53.	Leaves all basal	*A. kandaharica*
53a.	Leaves not all basal	*A. rhodocentra*
54.	Auricles very narrow	55
54a.	Auricles up to 1/2 achenes	57
55.	Involucral bracts ovato-oblong, hyaline-margined	*A. susiana*
55a.	Involucral bracts lanceolate	56

56.   Tube of disc florets inflated; ca. 2 x 0,5 mm achenes (very slender)
                                                                              A. gilanica
56a. Tube of disc florets not inflated; achenes up to 1,5 mm long *A. breviradiata*
57.   Plant appressed-pubescent, stems purplish. Auricles longer than 1 mm
                                                                              A. peregrina
57a. Plant tomentose-lanate. Auricles up to 1 mm   *A. tomentosa*
58.   Leaves subpinnatisect to pinnatisect   59
58a. Leaves 2-pinnatisect to 3-pinnatisect   61
59.   Leaves petiolate   60
59a. Leaves sessile; stems all foliated   *A. micrantha*
60.   Leaves all basal   *A. gayana*
60a. Stems all foliated   *A. austroiranica*
61.   Ligules up to 10 mm long   62
61a. Ligules more than 12 mm long   64
62.   Stems up to 15 cm   63
62a. Stems more than 15 cm   *A. leucolepis*
63.   Stems 4-8 cm; leaves 1,5 – 2,5 cm long; receptacle conical   *A. gillettii*
63a. Stems 10-15 cm; leaves 2 – 3 cm long; receptacle ovoideo-hemisphericum
                                                                              A. hirtella
64.   Internal involucral bracts becoming membranous; external involucral bracts
      dorsally wooly-hairy; leaves 2 – 3 cm long   *A. hirtella*
64a. All bracts equal; external involucral bracts green; leaves 1 – 2,5 cm long
                                                                              A. microlepis
65.   Involucral bracts dark-brown or brown margined   66
65a. Involucral bracts hyaline margined   68
66.   Involucral bracts subglabrous, somewhat woody at base   *A. anthemiformis*
66a. Involucral bracts pubescent   67
67.   Receptacle convex; basal tube of ray florets villous; basal part of disc
      florets not becoming inflated   *A. melanacme*
67a. Receptacle conical at maturity; basal tube of ray florets glandular; basal
      part of disc florets becoming inflated   *A. zaianica*
68.   Achenes with auricle or crown, at least in the outer achenes   69
68a. Achenes without auricle or crown, at least in the outer achenes   75
69.   Auricles 1/2 achene or more, all achenes auricled   70
69a. Auricles shorter as 1/2 achene, or only a crown   71
70.   Ligules c. 13 mm long; receptacle conical, very acute   *A. scopulorum*
70a. Ligules up to 9 mm long; receptacle hemispherical   *A. moghanica*
71.   Achenes up to 2 mm long   72
71a. Achenes longer than 2 mm   73
72.   Involucre villous; leaves sessile   *A. hamrinensis*
72a. Involucre glabrous; leaves petiolate; very slender achenes   *A. rascheyana*
73.   Ligules up to 8 mm long   74
73a. Ligules c. 12 mm long   *A. brachystephana*
74.   Involucral bracts ovate-oblong; pales mucronate. Base of disc florets
      purplish   *A. hamrinensis*
74a. Involucral bracts lanceolate; pales acuminate.   *A. wallii*
75.   Achenes truncated at apex   76

75a.	Achenes rounded at apex	*A. microsperma*
76.	External involucral bracts lanceolate; receptacle conical	*A. candidissima*
76a.	External involucral bracts ovate; receptacle hemispherica	*A. plebeia*
77.	Involucral bracts light brown to dark-brown or dark margined	78
77a.	Involucral bracts pale or hyaline margined	89
78.	Disc corollas becoming inflated at maturity	79
78a.	Disc corollas not becoming inflating at maturity	*A. hyalina*
79.	Auricles of ray florets as long as achene; involucre glabrous	*A. chia*
79a.	Auricles up to 1/2 achene or absent; involucre glabrous or villous	80
80.	Involucre glabrous	81
80a.	Involucre sparsely to densely hairy	83
81.	Pales caducous at maturity	*A. monilicostata*
81a.	Pales persistent at maturity	82
82.	Auricle absent	*A. confusa*
82a.	Auricle present but only in the disc achenes	*A. secundiramea*
83.	At least outer achenes with auricles	84
83a.	At least outer achenes without auricles	86
84.	Auricles 1/2 achene	*A. parnesia*
84a.	Auricles very short	85
85.	Stems basally tinged with red	*A. ubensis*
85a.	Stems not so	*A. lorestanica*
86.	Leaves glandular-punctate	87
86a.	Leaves not glandular-punctate	*A. arvensis*
87.	At least achenes of peripheral disc florets tuberculate	88
87a.	Achenes smooth	*A. taubertii*
88.	Auricle absent	*A. confusa*
88a.	Auricle present but only in the disc achenes	*A. secundiramea*
89.	Achenes with auricle or crown, at least only outer achenes or only inner ones	
		90
89a.	Achenes without auricle or crown at all	109
90.	Auricles as long as 1/2 – 1/3 achene or more	91
90a.	Auricles up to 1/4 achene or only a crown	99
91.	Leaves fleshy	*A. aeolica*
91a.	Leaves not fleshy	92
92.	Stem purplish at base	*A. busherica*
92a.	Stem not purplish at base	93
93.	Peduncle clavate at maturity	94
93a.	Peduncle not clavate at maturity	98
94.	Leaves 2-3 pinnatisect	95
94a.	Leaves pinnatisect to pinnatifid	*A. werneri*
95.	Involucre glabrous	*A. tricornis*
95a.	Involucre villous	96
96.	Auricle strongly fimbriate	*A. schizostephana*
96a.	Auricles intere or breviter dentate	97
97.	Achenes slender, 2 x 0,15 mm; involucre 5 – 7 mm in diameter	
		A. hemistephana

97a. Achenes cylindrical c. 2 mm long but not slender; involucre 10 – 15 mm
     in diameter   *A. auriculata*
98. Involucre glabrous; leaves glandular-punctate   *A. macedonica*
98a. Involucre villous   *A. arenicola*
99. Auricle 1/5 – 1/4 achene   100
99a. Anguste crown or auricle   104
100. Peduncle clavate at maturity   101
100a. Peduncle not clavate at maturity   103
101. Achenes with ribs   102
101a. Achenes without ribs   *A. fimbriata*
102. Pales up to 2,5 mm, mucronulate; tube of ray and disc florets villous
                                                A. kurdica
102a. Pales longer than 2,5 mm, carinate   *A. atropatana*
103. Leaves glandular-punctate; achenes 1 – 1,75 mm long *A. macedonica*
103a. Achenes c. 2 mm long   *A. atropatana*
104. Peduncles clavate or slightly clavate at maturity   105
104a. Peduncles always not clavate at maturity   *A. macedonica*
105. Disc corollas becoming inflated at maturity   106
105a. Disc corollas not becoming inflated at maturity   *A. glareosa*
106. Ligules up to 3 x 2 mm   *A. werneri*
106a. Ligules longer than 4 mm   107
107. At least disc achenes ribbed and tuberculate   *A. glareosa*
107a. Achenes without ribs, almost with a short crown   108
108. Achenes up t 2 mm, outer ones acutely trigonal *A. leucanthemifolia*
108a. Achenes longer than 2 mm, obconical-prismatic   *A. emasensis*
109. Involucre glabrous   110
109a. Involucre villous   113
110. Middle and upper leaves sessile   *A. leptophylla*
110a. All leaves petiolate   111
111. Tube of disc floret inflated at maturity   112
111a. Tube of disc floret not inflated at maturity   *A. aaronshonii*
112. Leaves entired and mucronate or tripartite in the distal part *A. indurata*
112a. Leaves at least pinnatipartite   *A. arvensis*
113. Apical part of involucral bracts brown   *A. mauritiana*
113a. Apical part of involucral bracts as coloured as the margin   114
114. Peduncles remaining slender at maturity   115
114a. Peduncles clavate at maturity   117
115. Achenes not more than 1/2 as wide as long   116
115a. Achenes stout, more or less as long as wide, at least the outer ones
                                               A. arvensis
116. Stems up to 15 cm; leaves sub-2.pinnatisect to 2 pinnatisect; external
     involucral bracts triangular; tube of disc florets not inflated at maturity
                                         A. tenuicarpa
116a. Stems longer than 15 cm; leaves pinnatifid or pinnatisect; external
     involucral bracts acute; tube of disc florets strongly inflated at maturity
                                         A. ruthenica
117. Achenes inconspicuously striate, stem pubescent   *A. ammophila*

117a. Achenes striate   118
118. Distal part of disc florets funnel-shaped; tube of disc florets usually inflated
      at maturity   *A. arvensis*
118a. Distal part of disc florets not funnel-shaped; tube of disc florets not inflated
      at maturity   *A. aaronshonii*
119. Ligules yellow   120
119a. Ligules white, pinkish or absent   125
120. Ligules up to 7,5 mm long   121
120a. Ligules longer than 7,5 mm   124
121. Involucral bracts ovato-lanceolate   122
121a. Involucral bracts triangular, at least the outer ones   *A. marschalliana*
122. Lower leaves with c. 12 pairs of primary segments, each divided into c. 7
      linear crowded secondary segments   *A. marschalliana*
122a. Lower leaves with 3-4 pairs of primary segments, each divided into
      c. 3 oblanceolate or obovate secondary segments   123
123. Involucre generally less than 1 cm broad; upper leaves pinnately lobed
                                                                                     A. kotschyana
123a. Involucre 1 - 1,5 cm broad; upper leaves cuneate, entire or shallowly
      lobed at apex   *A. pauciloba*
124. Stems 9 – 16 (-30) cm, ascending, usually simple; ligules 3,5 – 6 mm
      wide; achenes c. 2 mm, not granulate; young leaves densely lanate
                                                                                     A. trotzkiana
124a. Stems 40 – 60 cm, erect, corymbosely branched; ligules up to 3 mm
      wide; achenes 1 – 1,5 mm, granulate, mainly on the angles; leaves
      sparsely hairy   *A. virescens*
125. Ligules pink, at least at base   126
125a. Ligules white or absent   128
126. Receptacular scales purplish-pink mainly at the apex; leaves lobes
      mucronate   127
126a. Receptacular scales not so; leaves lobes not mucronate   *A. tricolor*
127. Basal leaves dull green; achenes longer than 2mm, distinctly ribbed.
      Endemic to Sicily   *A. aetnensis*
127a. Basal leaves sericeous-pilose; achenes up to 2 mm, obscurely
      4-angled. Endemic to Cyprus   *A. plutonia*
128. Ligules present   129
128a. Ligules absent   156
129. At least the inner achenes tuberculate   130
129a. Achenes not tuberculate   135
130. Corolla tube of disc florets inflated at maturity   131
130a. Corolla tube of disc florets not inflated at maturity,
      leaves glandular-punctat   132
131. All achenes with prominent ribs   *A. ismelia*
131a. At least inner achenes without ribs or slightly ribbed   *A. cretica*
132. Receptacular scales equalling the florets   *A. ismelia*
132a. Receptacular scales shorter than the florets   133
133. Capitula (6-)20-30(-38) mm in diameter; involucres (5-)9-13(-14) mm in
      diameter; inner involucral bracts (3,1-)3,4-5,3(-6) mm long; ray florets

(5,9-)7,6-12,9(-15,7) mm long; disc florets (1,3-)1,6-2,1(-2,6) mm long; pales (2,6-)2,9-4,3(4,6) mm long; achenes (1,3-)1,6-2,1(-2,6) mm long

A. pedunculata

133a. Capitula (30-)35-47(-55) mm in diameter; involucres usually (10-)14-22 mm in diameter; inner involucral bracts (4,5-)5,4-7,5(-8,2) mm long; ray florets (11,0-)11,8-18,8(-25,5) mm long; disc florets 3,1-3,8(-4,2) mm long; pales (3,7-)4,1-5,5(5,8) mm long; achenes (1,8-)2,0-2,9(-3,1) mm long 134

134.   Achenes always without corona   *A. abylea*

134a. Achenes with an adaxial corona   *A. punctata*

135.   Outer and inner involucral bracts of the same form   136

135a. Outer and inner involucral bracts of different form   144

136.   At least upper cauline leaves lobulate or entire   137

136a. Leaves 1-2 pinnatisect   139

137.   Margin of involucral bracts brown   138

137a. Margin of involucral bracts hyaline; capitula 2,0 – 2,5 cm in diameter

A. argyrophylla

138.   Plant white-pubescent; receptacle conic.   *A. didymaea*

138a. Plant subglabrous or sparsely pubescent; receptacle triangular. Capitula 4 – 5 cm in diameter   *A. cretica (subsp. iberica)*

139.   Involucral bracts with dark or ± narrow pale-brown margin   140

139a. Involucral bracts not different at margin   *A. jordanovii*

140.   Involucral bracts with dark margin   *A. cretica*

140a. Involucral bracts with pale-brown margin   141

141.   Achenes of inner florets not or inconspicuously ribbed   142

141a. Achenes of inner florets ribbed   143

142.   Leaves glandular-punctate   *A. cretica*

142a. Leaves not glandular-punctate   *A. pindicola*

143.   Ligules 6 mm long; corona absent or minute; involucre white-tomentose

A. calcarea

143a. Ligules 10 mm long; corona ± entire, c. 0,25 mm (c. 1/7 achenes); involucre glandular-pubescent   *A. acyphylla*

144.   Involucral bracts with black or dark-brown margin   145

144a. Involucral bracts with pale to brownish or hyaline margin   148

145.   Leaves glandular-punctate   146

145a. Leaves not glandular-punctate   147

146.   Stems (20-)30-60 cm tall; capitula up to 6 cm   *A. cupaniana*

146a. Stems up to 22 cm   *A. meteorica*

147.   Cauline leaves short-petioled; pales linear-spathulate. Endemic to Caucasus

A. cretica

147a. Cauline leaves sessiles; pales oval, mucronate. Endemic to Lebanon

A. didymaea

148.   Leaves petiolate   149

148a. Leaves sessile   *A. orientalis*

149.   Capitula up to 25 (30) mm   150

149a. Capitula 30-40 mm   *A. tranzscheliana*

150.   Leaves fleshy   *A. marittima*

150a. Leaves not fleshy   151

151. Involucre glabrous   *A. meteorica*
151a. Involucre ± hairy   152
152. Achenes tethraedrical, without auricle; leaves densely white-pubescent
                                                                     A. fruticulosa
152a. Achenes not tethraedrical, with or without auricle   153
153. Stems up to 20 cm   154
153a. Stems 20 – 40 (-50) cm; leaves somewhat fleshy
                                                          A. cretica (subsp. gerardiana)
154. Auricle present   155
154a. Achenes with only a narrow crown   *A. regis-boriis*
155. Receptacle conical-acute, more or less hairy   *A. rumelica*
155a. Receptacle hemispherical, silky-pubescent   *A. sterilis*
156. Leaf segments linear-acuminate, acicular   157
156a. Leaf segments oblanceolate or obovate   159
157. Plants suffrutescent (Is. Rodhos)   *A. rhodensis*
157a. Plants herbaceous or only slightly woody at base   158
158. Capitula borne on ± leafy stems; involucre up to 1,5 cm broad; receptacle
      ± conical; outer phyllaries lanceolate   *A. aciphylla*
158a. Capitula borne on leafless peduncles; involucre up to 1 cm broad;
      receptacle ± hemispherical; outer phyllaries ovate   *A. orientalis*
159. Margin of involucral bracts brown or pale-brown   160
159a. Margin of involucral bracts hyaline or membranous   163
160. All involucral bracts equal, acute. Ligules usually absent   *A. alpestris*
160a. Inner and outer involucral bracts different   161
161. Capitula campanulate, ± turbinate at base; inner involucral bracts
      with broad (c. 2 mm) obtuse scarious apices   *A. calcarea*
161a. Capitula hemispherical, rounded or umbilicate at base; inner involucral
      bracts with narrow acute apices   162
162. Involucre 0,7 cm broad; achenes 1,5 mm long; pales shortly pointed;
      rounded hemispherical shrublet.   *A. xylopoda*
162a. Involucre 0,75 – 1,5 cm broad; achenes 1,5 – 3,5 mm long   *A. cretica*
163. At least upper leaves entire   164
163a. Leaves pinnatisect   165
164. Stems c. 30 cm tall; leaves obtusely toothed or lobed   *A. pauciloba*
164a. Stems 15-25 cm tall; leaves with fine acute teeth   *A. cuneata*
165. Achenes distinctly ribbed, with shorter auricle   166
165a. Achenes not or only slightly ribbed, at least the inner ones   167
166. Leaves densely sericeous-tomentose   *A. cretica (subsp. spruneri)*
166a. Leaves glabrous or sparsely hairy   *A. pedunculata*
167. Achenes 2 – 2,5 mm long   168
167a. Achenes shorter, up to 1,75 mm long   170
168. Receptacular scales dentate or lacerate at apex; sometimes brown at apex
                                                                        A. cretica
168a. Receptacular scales truncated at apex, mucronate, linear   169
169. Receptacular scales acuminate; disc florets not or only slightly inflated
      at maturity   *A. kotschyana*

169a. Receptacular scales linear-oblanceolate; flowers strongly inflated at maturity
     A. cretica
170. Achenes up to 1,5 mm long   171
170a. Achenes more than 1,5 mm long, ± squarish in section   *A. cretica*
171. Stems up to 15 cm; leaves up to 2,2 (-3,5) cm long; achenes 1,25 (-1,4)
      mm long   *A. abrotanifolia*
171a. Stems 25 – 45 cm; leaves up to 4 cm long; achenes 1,5 mm long
      A. hydruntina
172. Achenes distinctly dilated at base; involucral bracts and receptacular
      scales falling off at maturity   173
172a. Achenes not dilated at base   → **Sect. Maruta**
173. Stems up to 10 cm   *A. odontostephana*
173a. Stems 10-25 cm   *A. tubicina*

Sect. Maruta (modified from Yavin, 1970)

1. Receptacular pales absent   *A. macrotis*
1a. Receptacular pales present   2
2. Plants annual   3
2a. Plants perennial   18
3. Mature fruiting heads very small, 5-8 mm in diam   4
3a. Mature fruiting heads exceeding 10 mm in diam   5
4. Achenes without ribs   *A. hermonis*
4a. Achenes ribbed   7
5. Involucral bracts scarious-margined   8
5a. Involucral bracts hyaline-margined   6
6. Disc corollas becoming inflated at maturity   *A. bornmuelleri*
6a. Disc corollas not becoming inflated at maturity   *A. pungens*
7. Achenes slightly curved, white, with 4-5 slightly winged ribs; auricle, short,
      cut into narrow lobes. Plants of sandy and salty deserts in Central Asia.
      A. deserticola
7a. Achenes straight, brownish, with rounded ribs, bald, somewhat rounded
      at tip. Plants of marshes, river banks and irrigated fields becoming very dry
      in summer.   *A. microcephala*
8. Receptacle hemispherical or broadly conical, covered all over with pales. 9
8a. Receptacle ovate or oblong-conical, bearing pales in the upper part only. 10
9. Involucral bracts usually dark-margined. Achenes shallowly
      sulcate, subtuberculate, bald. Desert plants of sandy habitats in Egypt.
      A. retusa
9a. Involucral bracts white-margined. Achenes deeply sulcate, smooth or
      rarely subtuberculate, with long or a short auricle, or tipped with 1-4
      minute teeth or entirely bald. Non-desert plants.   *A. galilaea*
10. Receptacle ovate. Achenes persistent at maturity.   11
10a. Receptacle oblong-conical. Achenes persistent or falling off easily at
      maturity.   12

11. Stems branching from above the base, moderately spreading. Leaves pinnatisect. Ray florets fertile. Achenes costate, somewhat muricate, with a lobed-dentate crown.   *A. adonidifolia*

11a. Stems branching from base, very widely diffuse. Leaves bipinnatisect. Ray florets sterile. Achenes costate striate, rough-tuberculate, with a short, one-sided margin.   *A. patentissima*

12. Achenes persistent at maturity. Ray florets fertile.   13

12a. Achenes falling off easily at maturity. Ray florets sterile.   16

13. Peduncles distinctly thickened at maturity. Outer achenes very thick, nearly quadrangular.   14

13a. Peduncles only slightly thickened under the head in fruit. Outer achenes not thickened.   15

14. The thick peduncles usually strongly bent in fruit. Auricle triangular or lobed, half as long as the achene or shorter; sometimes achenes with no true auricle, truncate or rounded at apex.   *A. pseudocotula*

14a. The thick peduncles remain erect in fruit. Auricle entire or lobed, often as long as the achene and forming a right angle with it.   *A. tripolitana*

15. Achenes obpyramidal, whitish, fungose, costate and smooth, rounded at apex, with a short inflated margin. Plants of sandy places in S.W. Iran. *A. fungosa*

15a. Achenes obconical, sulcate and prominently tuberculate, truncate at apex, with a very short one-sided margin. Plants of heavy, alluvial soils in the Mediterranean region.   *A. parvifolia*

16. Leaves finely bipinnatisect, the lobes cut into long, narrow and hair-like lobules. Achenes lightly costate, rather smooth, bald. Plants of ruderal places in Lithuania.   *A. lithuanica*

16a. Leaves finely bipinnatisect, the lobes cut into shorter, narrow-linear, mucronate lobules. Achenes more or less tubercled, bald or auricled.   17

17. Achenes sulcate and tuberculate; apex rounded, bald, appearing somewhat crenulated owing to the truncate ribs.   *A. cotula*

17a. Achenes sulcate-angulate, subtuberculate, with a thin, sharp, unilateral crown which is broader in the inner achenes. Endemic in S. Turkey on sandy river banks.   *A. corymbulosa*

18. Stem white-pubescent; leaves sessiles, grey-green. Achenes cylindrical, 2 x 0,8-1,5 mm   *A. tigrensis*

18a. Stem sparse hairy; leaves short-petiolate, laxly hairy to glabrous. Achenes tetragonous, 1-1,5 x 0,75 mm   *A. yemensis*